Counting the Floats

A good start !!!

Floats	0.1	0.2	0.3	0.4	0.5	0.6	0.7	0.8	0.9	1.1
Counters	1	2	3	4	5	6	7	8	9	10

Making progress ...

Float of $\approx \pi$	3.14159265358979323846264338
Counter	2532743338823081391461637905

(We can do this forever and beyond...)

by

Tamas Varhegyi

Author of Applied Conics, © 2006

June 4, 2018

I dedicate this book to :

My dear parents who - after two World Wars and a revolution –
had the courage and love to raise me and my sisters.

My wonderful children and grandchildren. I am really proud of them.

The first-rate medical professionals in Monroe, NC who take excellent care of me.

Many thanks for Lori Williams and her company Bar Code Graphics Inc. for providing
invaluable assistance in preparing this book for publication.

Last but not least, I am indebted to Waterloo Maple Inc. whose first rate algebraic
software tool, Maplesoft I used to design and implement the flagship algorithms
presented in this book. Most of my work for the past 25 years, including my
first book was made possible only by using their tools.

No man has earned the right to intellectual ambition until he has learned to lay his course by a star which he has never seen—to dig by the divining rod for springs which he may never reach.... Make your study heroic, for to think great thoughts you must be heroes as well as idealists.

Only when you have worked alone—when you have felt around you a black gulf of solitude more isolating than that which surrounds the dying man, and in hope and in despair have trusted to your own unshaken will—then only will you have achieved.

Thus only can you gain the secret isolated joy of the thinker, who knows that, a hundred years after he is dead and forgotten, men who have never heard of him will be moving to the measure of his thought – the subtle rapture of a postponed power, which the world knows not because it has no external trappings, but which to his prophetic vision is more real than that which commands an army."

Justice Oliver Wendell Holmes, Jr.
[From a lecture to Harvard students 1886]

Table of Contents

Introduction

Chapter 1

Infinity has no place in computations. Use omega [ω] instead

Chapter 2

Cantor's Diagonal Argument

Chapter 3a

Algorithms versus integers and floats

Chapter 3b

Structure and syntax rules of integers and floats

Chapter 4a

**Counting agents, countability, enumeration
Boneheaded counting**

Chapter 4b

Value and string length divergence for floats

Chapter 5a

**Complete Tree Structures Definition
One-to-one Correspondence between base-11 integers and base-10 floats**

Chapter 5b

**Complete Tree Structures Implementation
Generating candidate floats**

Cornerstone definitions and rules for this book :

Positive integers are the universal counting agents both for enumeration and counting.

In this book, all counting objects will be a well-defined subset of **floating point numbers**. We will refer to them with the abbreviation "**floats**".

Enumeration is the act of obtaining the total number of items in a **finite set** by counting. The counting is performed by open outcry and produces a single positive integer value.

Infinite sequences cannot be enumerated but they may be **countable**.

Counting is an act of associating a unique counted number (a float) with a unique positive integer. The association must be predictable, computable and permanent.

Non-finite sequences of numbers which are generated according to predictable rules are countable. The rules are paramount as the one-to-one association is not possible without knowing them.

The natural existence of the predictable association between the floats and the positive integers enables us to develop a pair of two-way algorithms :

    1. From a know float we can generate a unique positive integer.

    2. From a unique positive integer we can retrieve the associated float.

Algorithms are an open ended stream of rules which generate either numbers or other entities : musical notes, colors, poems, shapes, drawings, 3D-sculptures, assembly instructions and so on. They are the product of intelligent agency and as such can never be predicted, bound or counted.

Open-ended collection of numbers which are generated using arbitrary rules and allow an unlimited number of duplicate floats are not countable. .

Another ways to look at this : There are an unlimited and unpredictable number of ways any given float can be produced by various algorithms. The float outputs of a random collection of algorithms which land on the same location on the number line are not countable.

Example : 3.14 is universally known as the first approximation of Pi.
But the floats in the list [6.28 / 2, 0.314 * 10, 20 – 16.86, SQRT(9.8596)] also produce 3.14

Note : We are stating the obvious when we notice that the positive integer counting agents are self-counted by definition.

Please see Chapter 4a and of course the entire book for more details !!

Introduction

Background :

1. The confusing state of set theory.
2. The unproven Cantor Diagonal Argument (CDA)
3. Boneheaded counting scenarios misapplied in proofs
4. Confusing algorithms with floating point numbers
5. Questionable claims about relative sizes of integer and float sets
6. Unprovable claim stating that floats cannot be put into 1:1 correspondence with integers

Project goals reached :

1. Clarify all related set-theory statements and disprove false arguments
2. Prove that floats are countable
3. Count the "uncountable" floats using representation size first, value second
4. To count floats between two given limits
5. To assign unique integer sequence numbers to each and every possible float
6. To retrieve any float in an arbitrary order, given its integer sequence number

History - how did this project start :

Over the years I have occasionally amused myself with the mathematical riddles, paradoxes and challenges which are abundant on YouTube. There is a sample list of claims I had problems with :

1. The unproven Cantor's Diagonal Argument **(CDA)**
2. Misapplication of infinity used to present fake proofs e.g. : $1+2+3+... = -1/12$
3. Confusing dynamic algorithms with static floats (e.g. π is not a float !)
4. Poorly defined syntax for floats (leading/trailing zeros or decimal points and so on)
5. Proliferation of tunnel-vision or boneheaded counting to prove uncountability of floats
6. Wrong claims about comparing number of floats in specific ranges to number of integers
   (e.g. "There are more floats between 0 and 1 than there are positive integer counters)
7. The mistaken belief that floats cannot be 1:1 correspondence with integers
8. Statements like this : "Because **we can never line up all the real numbers in a particular order,** we cannot establish a one-to-one correspondence between the set of real numbers and the set of natural numbers, and therefore the cardinality of the set of real numbers cannot be $\aleph_0$. The cardinality of the set of real numbers must be a different infinite "number!" Another Infinite "Number": $\aleph_1$

It was a slowly accumulating suspicion the mathematics of basic set theory and limits has gone off the rail at a number of junctions. Within a matter of weeks suspicions hardened into an unshakeable conviction. The discipline of set theory was due for a major correction, with respect to a number of questionable pivotal claims.

What do I bring to the subject matter :

Much needed clarity and simplicity.
The unique, powerful and versatile Complete Tree Structure (CTS)
A logical step-by-step analysis debunking the current state of flawed dogmas.

What has been accomplished :

Created a streamlined and unambiguous definition for the floating point numbers.

This new category of numbers we will call "floats" excludes all alternate, convenient forms and the entire class of algorithms. For sake of clarity, elegance and in an effort to maintain focus throughout the project, the treatment of the floats will be limited. Only decimal, positive, non-scientific, genuine, non-ambiguous and minimal character representations will be allowed.

Invented and designed a conversion algorithm which can convert every positive integer without limit into a positive float and every positive float into its corresponding positive integer at will.

A very unusual and novel non-monotonous, periodically rising and sinking sequence of floats was invented, which when input to a special algorithms will generate consecutive, positive integers in increasing order.

Designed a second algorithm which takes the positive integers just generated and produces the same sequence of floats which generated these positive integers in the first place.

It will be possible to count both ways, without leaving any gaps.

We are thoroughly familiar with the proper sequencing of the positive integers from the anchor value of zero. The cascading-carry method of counting is well know to any elementary school student.

We can even draw the corresponding - so called Complete Tree - which will house the entire progression of the positive integers. These integers may be retrieved by simply traversing the Complete Tree Structure and announcing the integers which are associated with each node.

But what was never done before is a way to count floats starting from an anchor value. Ask anybody to name the very first positive floating point number (float for short) and they will draw a blank. Make a guess yourself as you read it, you will agree that it is not a trivial question and many of you will respond that there is no such a thing as the first float !

In these pages you will learn how to start, proceed and associate each and every positive float with a corresponding sequential integer counting number or what we will frequently call counting agents.

To carry out this project, I attempted to unify the occasionally fragmented, confused and just plain incorrect ways dealing with float and integer cardinality and countability rules.

I will argue that there is no difference between the floats and integers with respect to handling the issue of cardinality and countability.

I will however make a strong case for a game changing distinction between algorithms and numbers.

You should agree with me that most of the countability issues had their roots in the merging and mixing of the wide-wild-world of algorithms with the much more orderly and organized set of real numbers.

Disclaimer : Although the term countability should strictly be applied to finite sets , in these pages I will use it to indicate that the members of a particular set can be paired with members of the set of positive integers. I experimented with introducing enumerability which neither the integers nor the floats possess, but it is a non-starter, so countability will have to do.

Application of the two-way conversion algorithms :

We might ask : What is the practical use of a pair of algorithms which can convert between two
representations of an endless progression of numbers? One, the set of positive integers the other,
a strictly defined subset of the floating point numbers (no algorithms, no negative numbers ,
no scientific notation, no loose syntax rules).

One obvious application comes in mind :
Since each float can be uniquely converted into a positive integer it becomes possible to store
all floats as integers. This even works for float values with scientific notation :
we simply would have to convert them to standard float syntax first, then to integers second.

What helped me to reach my goals :

1. A fairly well-functioning boneheaded detector and a tendency to ask and dig until I get an answer.

2. A gut feeling that the cardinality of floats and integers are comparable and that they can be paired.

3. I honored the majority of traditional notation and methods as long as they did not clash with the
 ultimate goals I set out to reach. Long-held ambiguous practices - no matter how popular – were
ignored. After all if I agreed and followed some widely accepted conclusions this work would not exist.

3. Visualization, diagrams of trees or distribution of numbers, of sample outputs

4. Several times I simply created novel equations by observing patterns on outputs whose source were
some algorithms but unrelated to the final formulas (integer – float association in the L&B pyramid.)

Future plans :

Present the results to the mathematical research, publishing and teaching community.

Actively promote the theories and the algorithms which implemented the counting tasks.

It is my hope that they will be accepted in future mathematical or engineering literature
and in daily practices as well.

Disclaimer :

This document is not perfect as it shows hand-to-hand combat with failing health and advanced age.
There are issues with organization, sequencing, disproportionate weight being given to the treatment
of subjects. Logic, basic readability and use of language are all wanting. The hand-drawn diagrams,
although convey the intended logic sufficiently, are amateurish.

But I had to make a crucial decision :
Either work toward a truly polished and professional document and risk never finishing it, or settle
for a rough draft which conveys the essence of the very important discovery and rely on the feedback
of the readers and the assistance of professional authors to create an improved edition.

Suggestions for improvement will be appreciated. If you undertake to re-write or append the treatment
of a particular subject, I will accommodate it in the next version with appropriate credit given.

Chapter 1

Infinity has no place in computations. Use omega [ω] instead

1. Introducing lower case omega [ω]

It will be a better choice to stand for the idea of a counter which grows without limits, although its resemblance to the infinity symbol is a mixed blessing (good because it would remind us to a number which never stops growing, but questionable because it would be still interpreted as simply a substitute symbol for infinity.)

It is definitely worth repeating : ω has absolutely nothing to do with infinity. ω is an actual numerical counter, while infinity is either an abomination, an admission of defeat, or a hallmark of defective philosophical thinking. No, it is not a brilliant idea, whose comprehension requires the insight of geniuses. If you embrace infinity and sanction its use then you simply pretend that you understand it and as such you expect to be immune to any challenge. But it is all a mirage. Please note : $\omega \neq 0.5 \infty$ (my attempt at humor)

None of which applies to us, so here is what we are going to do : Whatever proofs we want to complete or behaviors to demonstrate, we will be satisfied if they will work for Very Big Numbers, or ω's for short. How big ? Your call, provided that it is a value not a wish.

Then we must prove that they will also works for **ω+1**. Usually this is done by employing induction proofs or other means.

2. The Confusing state of the mathematics of set theory involving the concept of infinity

Although the set theory topics we will deal with have a wide reach and a weird mixture of fact and fantasy there will be one single overriding backbone which they share. This commonality centers around the idea of infinity. To be more specific an, unproven, unwarranted and frequently amateurish misapplications of the idea of infinity. It is now common practice to use infinity in equations, algorithms and proofs claiming that infinity is an actual quantity.

One of the best known example of the misuse of infinity is Cantor's Diagonal Argument. His argument was hopelessly flawed but very few voices opposed him, mostly because they mistakenly believed that perhaps they are not smart enough to understand a sophisticated argument put forth by a prominent mathematician.

3. Historical opposition to infinity

Henri Poincare, one of the greatest mathematicians in history, rejected Cantor's theory, noting: "Actual infinity does not exist. What we call infinite is only the endless possibility of creating new objects no matter how many exist already".

 Carl Friedrich Gauss's views on the subject can be paraphrased as: "Infinity is nothing more than a figure of speech which helps us talk about limits. The notion of a completed infinity doesn't belong in mathematics'. In other words, the only access we have to the infinite is through the notion of limits and hence, we must not treat infinite sets as if they have an existence exactly comparable to the existence of finite sets."

1

<u>Niels Abel</u> : "If you disregard the very simplest cases, there is in all of mathematics not a single infinite series whose sum has been rigorously determined. "

4. Improper use of infinity

Under no circumstances would we attempt to disprove any theorem or formula which contains the infinity sign or the unfinished +... construct. The plain act of attempting to manipulate such material would lend legitimacy to infinity as a valid mathematical entity. The only rational approach to infinity is to stay away, present proofs using finite variables and leave it at that.

4.1. Can't use infinity in any expressions or calculations involving non-terminating series. As soon as infinity is mentioned, reason stops and guessing / fantasizing commences. Most series are evaluated only up to a handful of terms anyway, so use a variable to do it (ω)

4.2 We have to settle for ω and what it stands for : a very big number. That exists. It can be tailored to physical and theoretical circumstances, it can be used in algorithms, formulas and various scientific literature. ω is our friend, we can call it as soon or well before we run out of ink or perseverance when we deal with phenomena which refuses to terminate.

For example we can say : "Let's count positive integers : 1, 2, 3, 4 and so on to infinity" , or make it more formal and customizable : 1, 2, 3, 4, ..., ω-1, ω. The phrase "so on" has no place in formulas but a sequence which terminates in actual numbers does.

4.3 Infinity is not and never was a practical or conceptual tool to conduct physical, mathematical even conceptual scientific quest into our world. Yes, it might be used to entertain us with paradoxes (Cantor, Hilbert's paradox of the Grand Hotel, Banach-Tarski, Hyperwebster, (which is a good example of sloppy multiple counting and making erroneous assumptions based on that) but it has no place in serious scientific or philosophical discussions.

5. Nonsense examples

5.1 Hilbert's Hotel paradox : It is an existential contradiction, in addition to the set theory one : Here is why : If there was a credible attempt to create endless hotel rooms we would not be sittings at two different computers, I typing, you reading. The material demands in a futile attempt to create such hotel chain would have already been completely exhausted all the matter in this and an endless number of other universes. Moreover, if it was to be a credible effort it must have been going on since before existence itself. I figure that anything which has a beginning already would have missed its chance to reach infinity. At any point in time a candidate infinite stream should be already half-way there, meaning that it extends into the past indefinitely. I am just messing around here..., I know that I make no sense – but neither does anybody else.

Another perspective is that the entire thought experiment is resting on a fictional scenario : the completion of an infinite number of hotel rooms. If such construction has ever been attempted it would have to be an ongoing affair, so the hotel manager would not have the luxury of manipulate an infinite number of rooms, only a finite ones.
Once we take that step back from the impossible the paradox disappears.
Try it ! Redo the argument see how it unfolds.

5.2 Infinity used in equations leads to gibberish

We can't add to, subtract from, multiply or divide by infinity.

If we allow it we would have nonsense statement like :
$\infty * \infty = \infty$, now divide both sides by ∞ we get $\infty = 1$

how much is $\infty - \infty$ is it perhaps 0 ? or $2 * \infty / \infty$?

Can we write $\infty = \infty$, $\infty/\infty = 1$, $\infty + 1 = \infty$, $\infty/0 = \infty^2$ or $1/0 = \infty$?

is $\infty - \infty = 0$ true ?   How much is ∞^∞ ? and so on.

They are not just wrong but meaningless incomprehensible gibberish.

I strongly recommend the we remove infinity from all formulas, expressions, series, equations, sums, proofs and computer algorithms.

5.3 We can manipulate sequences leading to nonsense...for example :

Proving the two inequalities : $\infty * \infty < \infty$ **and** $\infty < 1$... **both nonsense**

2^∞	3^∞	5^∞	...	p^k	...	∞^∞
2^k	3^k	5^k	...	p^5	...	∞^k
16	81	626	...	p^4	...	∞^4
8	27	125	...	p^3	...	∞^3
4	9	25	...	p^2	...	∞^2
2	3	5	...	p	...	∞

We have ∞ number of primes horizontally.
Next, each prime is raised from the first to the ∞-th power.
Such powers then number $\infty * \infty$
Yet, most of the integer counters will never be generated that way.
What do you say ? Is $\infty * \infty < \infty$? If yes, then let's divide each side
by ∞ to obtain $\infty < 1$. There is "∞-ly" many idiotic ∞ arithmetic for you

5.4 Ultimate YouTube outrage : 1+2+3+..+ = -1/12

Surfaced on YouTube some time ago. Somebody has proceeded to
prove that $1+2+3+... = -1/12$ and got over 6 million hits.

Luckily the author of the following YouTube video disproved what is obviously a nonsense.
See the link https://www.youtube.com/watch?v=YuIIjLr6vUA

6. My comments and personal preferences

6.1 $1+2+3+4+...$ is undefined, not equal to anything so the $=$ sign cannot be prefixed or appended to it . $1+2+3+4+...+n = n*(n+1)/2$ this is correct

6.2 Inserting equal signs between some characters is not a right, it is a privilege afforded to those who know what they are doing. The presence of the ∞ sign in any mathematical construct will invalidate that construct.

6.3 I am unqualified to write an essay about infinity other than that I have no use for it.

6.4 I became convinced that the authors of the entire body of mathematical literature are stabbing randomly through the curtain of ignorance when it comes to infinity. Anything goes, nothing needs to be proved and everybody appears to be an expert about an entity which does not even exist and can never be accessed.

6.5 I strongly recommend the we remove infinity from all formulas, expressions, series, equations, sums, proofs and computer algorithms.

6.6 The proper domain to discuss infinity belongs to philosophy, science fiction and mental institutions but not in mathematical literature.

6.7 See Appendix : Case Study #1 : Converging geometric series using ω instead of ∞

Chapter 2

Cantor's Diagonal Argument

1. Historical Background :

In the year 1891, an argument with the basic premise that integers are countable but floats are not, was published by Georg Cantor. It became known as Cantor's Diagonal Argument or CDA and has been credited to him even though it was actually discovered by Paul du Bois-Reymond.

One contemporary narrative of CDA is shown below in decimal rather than the original binary form :

Cantor's Diagonalization Proof: Suppose towards a contradiction that there is a bijection $f : \mathbb{N} \to \mathbb{R}[0, 1]$. Then, we can enumerate the infinite list as follows:

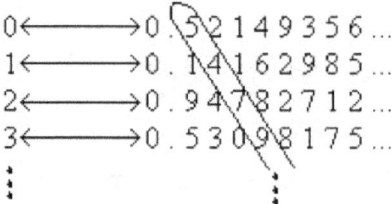

The number circled in the diagonal is some real number $r = 0.5479\ldots$, since it is an infinite decimal expansion. Now consider the real number s obtained by modifying every digit of r, say by replacing each digit d with $d + 2$ mod 10; thus in our example above, $s = 0.7691\ldots$. We claim that s does not occur in our infinite list of real numbers. Suppose for contradiction that it did, and that it was the nth number in the list. Then r and s differ in the nth digit: the nth digit of s is the nth digit of r plus 2 mod 10. So we have a real number s that is not in the range of f. But this contradicts the assertion that f is a bijection. Thus the real numbers are not countable.

That is it, see http://people.csail.mit.edu/alinush/math/countability.pdf for details

2. Current Status of CDA in circulation :

In the ensuing 6 generations - although it has never been upgraded to a proof - all challenges to CDA has been met with contempt often degenerating into name calling.

At the present it has been universally accepted as the undisputable proof that floats are uncountable.

After my first encounter with CDA I found myself intimidated by my inability to comprehend its logic, even though it appeared simple and straightforward. So I did what the literature dealing with infinity would consider a blasphemy. I modified CDA using finite values instead of a vague and confusing deployment of infinity (depicted by the nonsensical ellipsis [...] construct).

Where it was appropriate I substituted a tangible variable omega (ω) which stands in for an unlimited Very Big Number. Verbose, yes but it is very hard to ignore and leaves no interpretation to one's imagination.

Cantor made a mistake by not performing his calculations using actual variables such as a omega (ω) If he did that, the fatal flaw in his argument would have become obvious not only to him but to everybody reading his thesis. He went directly to the concept of infinity employing a sort of condescending nonsense ellipse notation. Smooth, nobody could ever get their hooks into something so "informal" and hazy.

It is important to state the significance of the "A" in the acronym. CDA is an "Argument", not a theory, and in spite of its universal acceptance it was never formally proven.

I have watched a number of YouTube videos, skimmed over the Wikipedia Cantor main article, then ventured into the Talk Pages on the subject. Soon came away with a strong suspicion that nobody really understands how CDA actually works or to what practical use it has. Then I noted that in the history of mathematics literature there was some mention that Cantor's contemporaries really disliked the diagonal argument, calling it metaphysical. That for me came as a surprise.

Today, a few serious mathematicians still refute CDA [Norman Wildberger] https://www.youtube.com/watch?v=XKy_VTBq0yk while others invite us to disregard Cantor's theory [Andrej Bauer]. But overall, the mathematical community celebrates it as one of most elegant truths in mathematics and Cantor's ideas are now taught unchallenged in all the universities of the world. Oddly, It even has a cult following. On some CDA forum hosted by Wikipedia, one proponent pigeonholed the CDA detractors as "truthers", whatever that connection might be.

This got my attention as I forever believed that mathematics was one branch of science which could not be corrupted or swayed by popular opinion. After all, if you are a mathematician and come up on a theorem, your main task is to attempt to prove or disprove them. Once there is a published proof, liking or disliking is not a proper response by any mathematician. CDA was not amenable for rigorous logical treatment, which is a good indication that it was outside of or contrary to the accumulated body of mathematical knowledge.

See also Appendix : Case Study #2 : Disproving Cantor's Diagonal Argument

Algorithms Versus Integers or Floats

Introduction

Of the 3 sets, **algorithms, integers and floats** by far the biggest is the vast world of algorithms. Yet they do not belong with integers or floats because :

1. They are instructions and not numbers.
2. They are undefinable both in variety and with respect to numerical outputs.
3. They can be generated by an arbitrarily large and open creative process : the human mind.
4. Thus the algorithmic output of a scientifically advanced race is open-ended and unpredictable.
5. I am not prohibiting or ignoring algorithms, I simply admit that they are uncountable.

Instead of counting floats a failed attempt was made to count the float outputs of a limitless collection of algorithms. Floats are eminently countable, because they are well defined. However, counting the output of all float-producing algorithms is impossible. Floats are not alone in this handicap. It is also not possible to count the outputs of integer-producing algorithms. For some reason, over the ages it was repeatedly attempted to count the outputs of an unlimited, disorganized and complicated group of algorithms which happen to produce floats. Not surprisingly all attempts failed outright.

It is not clear to me when these attempts to count float-producing algorithms mistakenly rendered the floats uncountable. As we will show, the simple character representation of the floats are eminently countable and even match the positive integers in a one-to-one correspondence. Putting the algorithms and floats into their own categories will make the floats universally countable and that will certainly up-end the prevailing wisdom on the subject.

The **counting-of-floats procedure** we created does one phenomenal trick, which is the creation of a one-to-one correspondence between floats and positive integers. We simply may consider the duo of (int_seqno, float_story) as universal, since they are tied together for all times to come.

1. Quick takes :

1.1 False claim : Floats are not countable because there are an endless number of algorithms which resolve to floats. The premise is certainly true, but the conclusion is not. We can generate an arbitrary number of algorithms which will produce the single float, 3.2 such as : 4.3 – 1.1, 4.4-2.2,……… n+3.2 – n and so on. Since this argument can be applied to any float, it would appear that the number of floats is hopelessly uncountable.
Response : There are also an infinite number of algorithms which resolve to integers, so any attempt to count algorithms which produce integers will fail. Still, we would not dream of re-classify integers as uncountable. So there is no link between the number of algorithms and countability of either integers or floats.

1.2 Current literature depicts number lines showing algorithms.
That is wrong ! Number lines were created to organize actual values of integers and floats. There is no place for algorithms which usually contain non-numeric instructions.

1.3 The float outputs of algorithms do not influence the number of floats at all.
Definitely true. [Ch 3a] The number of floats of a given size is a pre-determined constant

and is easily computable. (90 for 3-character floats, 1710 for 4-character floats and so on)

However, the usage counts of the receiving floats on the number line varies from 0 to an arbitrarily large number. Example : The float 0.1 can be generated by the algorithm :

$$\text{one\_tenth\_n} := \frac{10^n}{10^{n+1}} \quad \text{an endless number of times.}$$

1.4 The float outputs of algorithms are uncountable.

One objection might be that even though the character representation of the floats are countable, however the algorithms and their outputs which resolve to these floats are not.

Absolutely true. The supply of algorithms is so vast and so unpredictable that their outputs is simply not manageable at all. It is much worse! Even the algorithms themselves, regardless of their outputs are uncountable.

1.5 The integer outputs of algorithms are uncountable as well.
Definitely true.
There is no connection between the number of algorithms and the countability of their particular constructs (integers, floats) which receive their outputs. We don't have to try hard to convince ourselves that algorithms may just as easily generate an uncountable pile of integers as they do floats. Yet no one has ever doubted the countability of integers based on their association with the vast pile of algorithms. Then let's remain consistent and don't judge the countability of the floats based on their association with algorithms.

1.6 Integers are self-counting. We can successfully argue that most algorithmic outputs which produce floats can be converted into integer producing algorithms. Their outputs could not be counted either. Would that make the integers uncountable as well? Of course not I think not. Here is why. We only have to go : 1, 2, 3 and so on. Which is the short way of stating that integers are self-counting.

2. Algorithms related literature on the Internet

In Wikipedia https://en.wikipedia.org/wiki/Real_number another surprise was coming and I quote :
In mathematics, a **real number** is a value that represents a quantity along a line.
The real numbers include the fraction 4/3, and all the irrational numbers, such as sqrt(2) (1.41421356...,
the transcendental numbers, such as π (3.14159265...). Real numbers can be thought of as points on an infinitely long line called the number line or real line, where the points corresponding to integers are equally spaced. Any real number can be determined by a possibly infinite decimal representation, such as that of 8.632, where each consecutive digit is measured in units one tenth the size of the previous one.

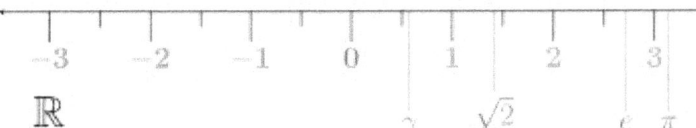

(note this is a picture copied from Wikipedia)

Look at this picture above it is showing $\sqrt{2}$, π , e, and so on. Also we are told numbers like 4/3 are also situated on this number line. Now it is true that there is a number line, and there are numbers on it, but those such as $\sqrt{2}$, π , e and 4/3 are not numbers they are algorithms. Can we still display their approximate values on the number line ?

Sure, if we label the particular points not with the name of the algorithm (which has no value and such has no definite location on the number line) but with one instance of its approximate value : So we can hang 1.33 on the number line but not 4/3. We can place 3.14 there too, but never π

I contacted them with the short message : "Hi, I would like to suggest an important correction to the number line you displayed. Please remove π, $\sqrt{2}$, , 4/3 as they are not numbers but algorithms. These algorithms have no definite values only approximations. The latter may be depicted on the number line, but algorithms which produce endless sequence of approximations have no home there." My suggestion was rejected unceremoniously and with a good dose of hostility and contempt. Personal issues notwithstanding, what amazed me was the absolute confidence and righteousness displayed by these individuals. This is usually a sure sign that the facts are not on their side. When I pressed them to give me the location of any of the algorithmic quantities on the number line this is what two of them gave me : **"Pi goes where pi goes."** End of discussion I reckon.

As the picture below shows I begged to differ.

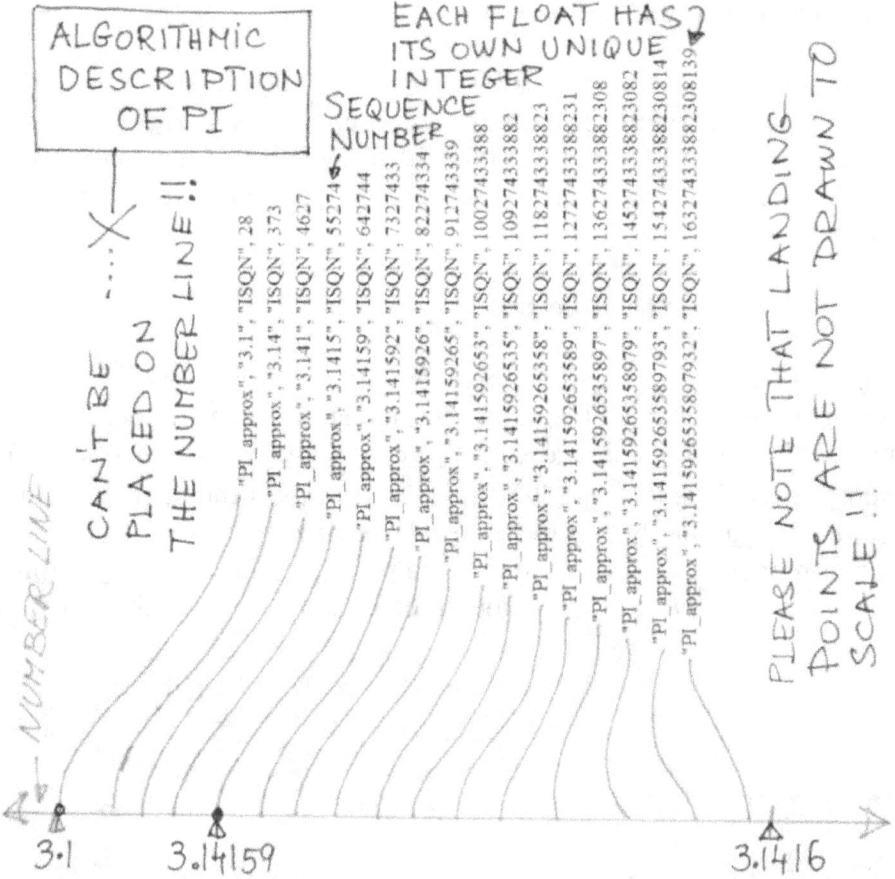

Demonstrating the positions of PI Approximations
on the float number line

Please note that none of the numerous algorithms which produce PI approximations belong on the number line. For a very good reason. Algorithms are an alphanumerical description of mathematical procedures thus cannot be represented on the number line which accepts only numbers. (That is why it is called the number line and not the algorithms line.)

I appreciated their empty response because it concurred with my earlier assessment that the "settled science" is completely wrong and that it will take more than one exchange to deliver this fact.

Of course that is nonsense : π is not on the number line, only its approximations are. The picture above demonstrates this clearly. According to Wikipedia, "the real numbers include all the <u>rational numbers</u>, such as the <u>integer</u> −5 and the <u>fraction</u> 4/3, and all the <u>irrational numbers</u>, such as sqrt(2) (1.41421356..., the <u>square root of 2</u>, an irrational <u>algebraic number</u>). Included within the irrationals are the <u>transcendental numbers</u>, such as π (3.14159265...). "

It is obvious from this definition that in addition to numbers, the definition of the real numbers include the wide open and undefinable cavalcade of algorithms (not only π, or sqrt(2), they are somewhat acceptable), but a lot more . For example x = f (a, b, c, ..., z), where f () stands for an arbitrary variety of algorithms which can be applied to an unlimited number and mix of input parameters.

I soon realized that nobody was counting floats which is a very simple and structured progression of digits (modified by a single power-of-base designator, the famous decimal point).

3. All attempts to count algorithms failed outright for many reasons :

3.1 Algorithms produce an endless variety of outputs, numerical, alphabetic, pictorial, musical notes, fractals and complicated design specifications.

3.2 The outputs may have an unlimited variety of forms and structures, for example :
Single value
Have a limited number of alternate values (roots of cubic equations)
Produce multi-valued unlimited number of approximations

3.3 Arbitrary numbers of algorithms may land on different and/or identical values on the number line.

3.4. There are an unlimited number of algorithms which can be created at will.

3.5. It is not possible to prevent widely disparate algorithms to produce duplicate floats.

3. 6. Attempting to obtain the float output of a single open-ended algorithm will appropriate all the theoretically available counting resources.

The simplest example would be the attempt to count the floats generated by the algorithm : 1 / 3. Yes, 1/3 is not a number : rather it is an algorithm which produces floats obtained by performing an endless succession of long-divisions : [0.3, 0.33, 0.333, ... 0.33333333....] Of course once we observe the pattern, we can toss the arithmetic steps of long-division and switch to the appearance clues : Start with the float 0.3 then append the digit 3 to obtain the next float. Don't stop...ever.

3.7 We can never devise a plan, as we are unable even to define the scope of the task. Each and every person may come up with an algorithm, or we can generate endless "algorithms which will in turn generate endless streams of floats. It is not the method, it is not the lack of resources. We simply have no idea how to do it. Imagine that you set out to build some road, then new specifications keep pouring in and the task just keeps expanding and ultimately never ends.

4. Properties of the outputs of algorithms:

There is a fundamental difference between well-behaved sets, infinite as they might be and sets which can't even be defined, described, bounded or comprehended, nor can they even be called sets in the first place. To say that examples are numerous is a gross understatement :
no matter how much we try to control them they will ooze out unbounded, uncontrollable.

For example : recursive generation of objects, infinitely stacked powers of multi-dimensional transcendental random sets, financial transaction records of an entire nation, positional coordinates of atoms in a universe or the cobwebs of infinitely expandable networks. The method of generating arbitrary sets, the random nature of their appearance or the disorganized way of processing the outputs puts them on the wrong side of possible.

4.1 Algorithms are vast, disorganized, incredibly diversified, resists any sane organizing attempts. A majority of them are uncountable. And they rule the world and our lives to a great extent. This might sound grandiose but if we go back to grade school, we can pinpoint the very first algorithms the students are exposed to multiplication, addition, subtraction and long division and so on.

In geometry the computation of length, perimeters, areas, volumes are all algorithms with the integer or float outputs being associated with the extents of tangible real life objects.

4.2 An overwhelming percentage of algorithms produce floats bridging the gap between theoretical physics, mathematics on one side and engineering, manufacturing everyday counting on the other. But algorithms are much more diverse infinitely large open-ended lot, producing everything from numbers to pictures, to musical compositions, chess game outcomes, weather and earthquake forecasting, Moon and planet faring trajectories and such.

4.3 Algorithms are chaotic while floats have strictly defined syntax. Great, we need both.

4.4 The false claim which links uncountability of floats to the vast undefinable heap of algorithms resolving to floats can be debunked by pointing out that there are an uncountable and undefinable number of algorithmic ways integers can be generated as well. If we allowed the counting of algorithms as a proof of non-countability for floats we would have to do likewise for the positive integers. Obviously, not being able to count the counting numbers is a contradiction. So we can safely abandon float producing algorithms as an argument for the uncountability of the floats.

4.5 For purposes of this project we will classify all positive decimal integers and floats as numbers. It is an arbitrary distinction made for the sole purpose of limiting the numbers to a very narrow but highly manageable set.

4.6 The rest of procedures, expressions and formulas will be potential candidates to be classified as algorithms, which then will do their specific work according to their design imperative.

4.7 The algorithms and the floats should never be mixed, confused, equated or in any way associated beyond a single one-way cause and effect relationship : In the course of their execution algorithms select one or more floats governed by the internal mathematical relationships of the algorithm. Thus floats are passive, always selected by some algorithm.

5.1 Multiple algorithms resolving to single floats or integers :

Allowing that kind of wide leeway of course leads to uncountability , not on its merit but because it is not possible to define the hopelessly open-ended lot of it. Here is one example : 0.1 is countable but the set of possible algorithms which can generate it is obviously open and uncountable. As you read this you can come up with dozens of algorithms each which will produce the float in a wide variety of ways. Examples : $k/(10k)$, $(k+n)/(10k+10n)$, $10^{(-1)} / 10^{(-2)}$ $2222/(11110+11110)$

Take for example the very obvious case of positive integers : 1, 2, 3, 4 and so on. Harmless enough, everybody can count them. Now let's expand the task and count **all the algorithms** whose numerical outputs are 3. After a few moments as the various ways 3 can be produced come into focus we all realize that it is utterly impossible not only to count, but far worse, even to define those algorithms. They are truly chaotic : e.g. $1+3-5+4$ or $165/55$ or $3^{15}/3^{14}$.

There is absolutely nothing above and beyond the float or integer representation as far as constant numbers go. We evaluate them and obtain a constant integer or float.

Note : As we place our first 4-character approximation of π on the number line we cannot state that 3.14 belongs to π. The reason is trivial : π is only one of an uncountable number of algorithms which will resolve to the float value of 3.14. The association is one way of course : The creators of algorithms have an "awareness" of the value distribution of the outputs on the float number line, but the reverse is not true. When we realized the idea of the number line with a particular layout extent and value distribution we had no idea which algorithms will use a given numerical position and even less comprehension what those algorithms actually do.

But it is a good division of functionality. Algorithms are the lifeblood of advanced sentient beings while the number line provides a permanent repository which is critical for the organization and sharing of the vast cavalcade of algorithmic outputs.

5.2 Examples of uncountable irrational or transcendental algorithms :

It is a different story with algorithms which do not and cannot produce closed forms of floats , for example : sqrt(2), π, 1/3, sin(x) , the golden ratio, Euler's constant and infinite fraction chains. Good point. They are categorically and irrevocably prohibited from being represented as a base-10 float string. Now I am asking for it, right ?

For what kind of math illiterate would suggest the exclusion of the incredibly important irrational and transcendental entities from their rightful place in the set of floats? I would, and for a good reason : Those famed mathematical icons are not NUMBERS, float, integer, transcendental or otherwise. Instead they must be classified as algorithms, containing instructions how to instantiate or convert the underlying equations into an instance of a float. So 1/3 becomes 0.333333…going on forever, sqrt (2) produces 1.4142136 in the first 8 digits.

All instantiations are approximations obtained by applying the algorithmic steps to a desired precision. The ever present famous π takes its place as the harmless 3.14 first, but then things get out of hand in the worst way. 3.14 is only the popular minimum length. Current estimates put the number of digits of the latest π approximation at over 10 trillion. Note :
If you want to start an argument at a party then ask what the value of π is. When everybody replies with 3.14, then contradict them by stating that it is actually 3, see what happens.

5.3 Algorithms generated by subverting the general flow of float counting (see Chapter 8)

Consider the sequence of 0.10101010....101010 ! Guess what. We just specified another algorithm on the fly, whose output is an infinite sequence of floats . But any set of instructions which generate float sequences contrary to the general execution flow of simple counting is an algorithm by definition. The reasons that such disruptors sabotage simple counting are :

1. The injection of a potentially infinite number of algorithms as a source of floats makes the counting progress itself unmanageable.
2. Each instruction neutralizes the efficiency of our "perfect" counting algorithm in some novel and unpredictable way.
3. Algorithms always generate huge customized gaps which generate sequences such that they potentially crisscross each other, or count sets multiple times. Such algorithms cannot be trusted.
4. Counting is not as much a mathematical operation but an organized task. We have to make sure that you have all the items to be counted available, sorted in some logical and unambiguous way with duplicates removed.

5.4 Examples of two fundamental floats algorithms :

At one level floats are numbers - an end in themselves but they can be used as inputs to an unlimited number of algorithms. Two of these algorithms are fundamental :

1. Convert floats to physical beads
2. Convert a float to its corresponding integer sequence number

Example invocations :
number_of_beads := convert_float_story_to_beads (1.47) will produce
number_of_beads = 147 individual beads each the size of 1/100 beads

int_seqno := find_int_seqno_from_float_story ("0.31") will give us
int_seqno = 118 counting integers

5.5 Examples of integer algorithms and sequences :

1. The sequence of $[2^k, 3^k,...n^k]$, with k going from 2 to VBN (Very Big Number)
2. The sequence which is obtained by listing the prime numbers and with each addition of a new prime proceed to generate all the possible products of the primes.

For example : 2*2, 2*3, 2*5, 3*3, 3*5, 5*5, 2*2*2, 2*2*3, 2*2*5, 2*3*3, 2*3*5, 2*5*5, 3*3*3, 3*3*5, 3*5*5, 5*5*5, and so on.

3. Use the products obtained in example 2 as exponents of all positive integers.
4. Any sequence in the form of : d, dd, ddd, dddd,...where d is any positive integer
5. Use the positive integers as anchors and from each start a new counting stream :
    1 – 2, 3, 4, 5, 6... then 2 – 1, 2, 3, 4, ... then 3 – 1, 2, 3, 4... and so on.

6. From each member of each endless sequence, hang off yet another stream, to form a kind of 3-dimensional progression.
7. Reciprocals of prime numbers
8. Factorials
9. $10^{(1(^2(^3)))}$....

10. Let us consider some of the ways the integer 1 can be generated : Let's list some of them : N-(N-1), N/N, N*(1/N), N^0, (1/N)^0, SQRT(N), where N is a positive integer from 1 to unlimited extent (of course we can use floats, negative numbers, expressions, provided that we evaluate them for proper domain and value (e.g. don't try SQRT(-5.7) or 0/0 etc.)

5.6 Custom made algorithms

1. Take for example any fraction with the 1/n where n, the denominator has at least one factor other than 2 or 5. For example 1/3, 1/7, 1/9, 1/(2*3), 1/(2*2*7) and so on. Such fractions do not have a closed form and cannot be counted. We can quickly agree that the set of multipliers in the denominator is simply undefinable. But let's not even get there.

2. Even without having any idea about the multitude of algorithms approximating π, we can all do our own : 3.13, 3.131, 3.1311, 3.13111 and so on. After a few trillion .111 digits let's change over to 3.1321... and go at it again. I have no idea how fast this converges to π, but if we ever end up with 3.15, we can just reverse and approach from above. After all an algorithm does not need to be efficient, nor easily understandable, it just has to make some half-baked effort at showing progress.

3. We can generate integer π's by simply removing the decimal point from each float approximation, to get 314, 3141 and so on. As a matter of fact the outputs of ALL float producing algorithms can be converted into integers. Yes it is quite possible that after the dp is removed that two unequal floats might resolve to the same integer, for example : 567.888 and 5.67888 both turn into integer 567888. So be it. The whole exercise is simply to show that as algorithms can produce both floats and integers, they outputs cannot be used to be flagship numbers for integer countability and float uncountability at the same time.
Note : All other floats which have the same or smaller representation size and a lesser value will have been assigned their own integer sequence numbers before the particular approximations of π obtained their own.

6. The float-counting procedures are themselves algorithms

It should not be a surprise to discover about the integer and floats that they themselves are the products of their respective algorithms. They are specific recipes codifying how to associate a computable number of physical objects with abstract number strings. Each and every digit in both integer and float number strings is associated with a power-of-base raised to a position-dependent exponent, which we might call the base multiplier. Once the product of the digit and the base multiplier is computed for each digit their sum is formed. This sum then can be associated with a corresponding number of beads of the proper size.

6a. Examples :

The integer 437 can be represented with 4 hundred + thirty + seven unity beads each with size 1. The float 3.9 corresponds to $3*10 + 9 = 39$ beads each one tenth of the size of the unity bead.

I am digressing somewhat , but the bottom line is that algorithms are at the mercy of our imagination and stamina and such are unlimited, unreliable while their outputs can be phenomenally useless (like π's after about 30 digits). But no matter how flamboyant they are, when the day is done, they must show themselves as a float, which is nothing more than a valid variation of an 11-character symbol set. There is no escaping from that straitjacket.

7. Usage count of floats and integers

Landing sites on the number line will be impacted multiple times by an undetermined number and mix of algorithms . Although a bare-bone number line is a dumb receptor without any knowledge about who and when accessed it, it is conceivable that some procedure may keep records of the algorithmic sources, the landing site float values and the total number of times that site was accessed by a mix of algorithms. It is easy to see that the records would be subject to alteration by an endless stream of algorithmic outputs. Unfortunately this state of affairs eventually was contributed to the nature of the floats themselves.

Float story	0.1	3.14	1.4142	0.333	0.3333	1003.44	0.0077	0.7	2000.4	0.9
Usage count	300	3439	7676	1111	307	59	4	1765	1	69007

If anybody attempts to use this pairing to claim that the floats are uncountable after all, there is a quick counterargument, shown in the imaginary table below : It catalogues the number of times the integer "1" was used last week on planet Earth. I doubt that anybody would seriously claim that the positive integers are uncountable based on the usage count presented. I rest my case.

Integer	1
Usage count	69007403403490230483098

Thus, if we adopt the rule of counting floats instead of counting the algorithms which produce floats, the counting task becomes as manageable as counting integers.

Each positive integer uses one or more characters selected from the 10-symbol decimal alphabet [0 1 2 3 4 5 6 7 8 9] Similarly a positive float is made up of one 3 or more characters selected from the 11-symbol decimal float alphabet [. 0 1 2 3 4 5 6 7 8 9] The particular makeup of each integer and float obeys a few common sense syntax rules. Floats are eminently and easily countable and are at near par with integers. In addition, in this chapter we will limit our discussion to algorithms which either produce valid floats or integers.

An easy conclusion may be reached that floats produced by the uncountable outputs of undefinable numerical algorithms are themselves uncountable as well (although very well defined at the moment when they are actually computed). However let's not lose hope just yet. It is without doubt that no matter how nebulous and arbitrary the floating point numerical output are , once one is recorded as a float it becomes permanent. This is possible only because the character length of each such output is set in stone. It is a trivial conclusion to state that the set of unique floats with a given character lengths is bounded.
For example there are only ninety 3-character or 2-digit floats. When we execute an infinite collection of algorithms, the output of any instance must be a float confined by its total character length. That fact has a universal consequence.

Let's say we examine the algorithms which produce 3-character floats. Guess what happens : No matter how large the set of such algorithms, their output is confined to 90 distinct floats. Which means that for each 3-character float we will have a potentially endless number of source algorithms. Did we count their unique outputs? You bet, it is the grand total of 90 floats. Did we count the algorithms themselves? Of course not, they are uncountable because they are undefinable. The number of duplicate outputs produced by the algorithms could be used as a tangible display of the overwhelming abundance of the algorithms when compared to simple counting numbers.

8. A confession and clarification :

1. Throughout this chapter I made a spirited case for the separation of algorithms and the two major categories of numbers. We insisted that both the integers and the floats are a frozen end product existing only for the sole purpose of organizing and storing the vast cavalcade of algorithms.

Yet anybody may have objected : The integers and floats are indeed receptacles but they are not the final inert sequence of numerical digits. Here is why. Both of them actually hold encoded instruction to generate the final product : a number of beads. So they are the last stage of algorithms before the rubber hits the road.

The integers will always generate the same sized unity beads, while the floats will have to generate ever so smaller bead sizes. But after everything is said and done, both will generate a bag of beads. Those bags are the final atomic product which cannot be dissected further.

2. One more observation may be made. Both the integers and the floats may be used for counting. Isn't that going to classify them yet another way to be essentially algorithms, like the conversion to beads is ? The answer is no. When we count we apply various algorithms (e.g. cascading carry) but these algorithms are applied to the integers and floats, which simply serve as inert arguments.

Conclusion :

1. We should not conclude that we are incapable of counting floats, but rather that the job is impossible if the specifications are vague and ever-changing.
Turns out that we are happy to accommodate a float-counting task which displays a rational and predictable schedule.

2. So, instead of dealing with the very liberal and exceptionally hard to work with world of algorithms we will limit ourselves to **positive integers and floats** which will be well defined and solidly anchored. The integer and float numbers are the link between algorithms and humans, they make civilized technological and mathematical progress feasible and organized.

3. I have no bone to pick with algorithms only with the misapplication of it : As a matter of fact it is my aim to complete the task of counting, which was long thought impossible. I offer an incredibly powerful companion algorithm to integer counting : the counting of floats. I hope it will be well-received.

4. Although its practical utility at this point is not entirely clear, the intellectual and moral worth is considerable. The proof that floats are countable will come as a shock as it has been declared impossible at least for the past 120 years. Countless (no pun intended) experts in the discipline of mathematics will have to revise the current literature and course materials. It will be an interesting time both for them and for the rest of us.

5. It is my goal that after the truly efficient float counting algorithms become well known that chapter of mathematical history will be behind us.

Chapter 3b

Structure and Syntax Rules of Integers and Floats

1a. Main theme : Restrict what is to be counted.

I coined a concise term which is not generally used in the mathematical literature : **Floats !**
Floats are a short name for the much longer floating point numbers and in our case
there are a lot more differences than just an abbreviated slang.

1b. The following entities will not be counted :

Float producing algorithms
Integer look-alikes
Negative floats
Scientific notation floats
Floats with illegal syntax
Non-decimal floats

1c. Floats are reduced the absolute minimum which is necessary to count them.

However once the counting operations are done, we can convert its string representation to the well-known floating point number format with all the attendant attributes, including the negative powers
of 10 association.

We can convert it to scientific notation, or a negative floating point number, assign it to variables and
use it in any arithmetic operation.

The reason for creating the bare-bone float data type was to achieve the ultimate simplicity and focus.
We wanted absolutely no incidental attributes which would have distracted us from the main show :
unambiguous ordering and conversion to their one and only integer counting agent.

1d. Counting negative and binary floats have been implemented. Also, scientific notation floats
certainly qualify by virtue of a simple two way conversion between them and regular floats.

But I will not showcase them in this manuscript. I judge it to be imperative to keep the focus on one
overriding task : the discovery and presentation of the fabulously designed **one-to-one, two-way,
float-to-integer** (and back) conversion algorithm.

That is what we all have been waiting for a very long time (well maybe some of us did !).
Hell, nobody did, because the overwhelming consensus was that **it simply could not be done.**

2. Definition of Floats :   (note that commas excluded from the character sequences)
Floats are specific character sequences which may contain one or more of the 11 symbols :
[**.** 0, 1, 2, 3, 4, 5, 6, 7, 8, 9] .

They belong to a single category of the **floating point numbers** which in turn are a subset of the
real numbers. In addition to the extremely spartan symbol set, floats must obey a severely
restricted set of **syntax rules (see below)**

Surprisingly, there is a very long list of mathematical constructs which will be excluded.
The reason for that is simple : I have embarked on an incredibly novel task of counting the floats

and an even harder quest to implement the universal and efficient generation of positive integer sequence numbers for each and every float possible. To make this goal achievable it was critical to define a robust and spartan syntax for the floats.

In the positive floats domain I have created a float definition which first accommodates every possible float string and second, does so exactly once. Alternate definitions which fall short on either of those 2 accounts will be summarily rejected as they don't fit into the goal of this project. However, I have no desire to advocate alternate definitions of the floats (although a removal of some ambiguous and unnecessary syntax rules could be beneficial), so please keep in mind that the rules I published are intended only for the scope of this project.

In retrospect the definition of floats which follows strict rules was a perfect fit, enabling the implementation of two complementary algorithms which establishes a truly simple one-to-one correspondence between floats and their corresponding integer sequence numbers. From the coding perspective, the tight specifications for the floats enabled me to complete the project in a reasonable time frame. Those, who frown upon the choices I made, would do well to remember :
An application which does everything and pleases everybody never gets written.

I would like to point out that if it becomes necessary, the implementation of more inclusive float definitions could be easily accomplished. For example, a front end which admits negative floats, maybe floats in scientific notation or binary floats (done that) would be some of the domains worth implementing. However that is an implementation detail and when done great care must be exercised as not to alter the general logic of the algorithms.

3. Design constraints :

1a. We will deal only with proper positive decimal integers (no leading zeros)
1b. Allow strictly character oriented reduced syntax floats only. See below for rules.
1c. The implied negative powers of 10 association for floats will be completely disregarded.
1d. The phrases "number of floats" or "number of integers" must be accompanied by
    sufficient limit specifications. If not, we will consider them meaningless.
1e. Obviously all algorithms must be excluded, as they are not constant numbers but
    written instructions or very complicated software procedures.

2. Integer counters can trivially count any set of integer numbers or count themselves.

3. The floats (once the implied powers of base are ignored) are nothing more than strings of numerical digits with an extra character, the decimal point. Thus, counting floats should be just as manageable as counting the integers provided that we have the discipline to enforce some basic syntax rules.

4. The overriding enabling design principle for the counting of floats is this : We must count **every float** and we must count each **exactly once**.

5a. Whatever string we designate as a candidate float must have a one-to-one correspondence between the string and the associated value of the float. That is why multi-valued outputs of algorithms are always rejected.
For example : If you ask anybody whether 1/3 is a float or not, 99% of them will say YES it is a float. That answer is wrong because 1/3 is an algorithm and not a float, with an endless number of approximate values, each of which is a valid float e.g. [0.3, 0.33,…, 0.33333…]. We obviously cannot classify 1/3 as a float as this would result in a float which would be capable of generating other floats. But that is not allowed as each float is defined as an inert end-product.

6. The reverse is also true : Each and every floating point value must be referred to by the simplest and shortest solitary string amongst the many possible strings which might resolve to that float value. For example since 3/10 resolves to a float value, we represent it by "0.3" and that is it. Alternates like "0.30" are not allowed.
Moreover this rule prohibits alternates like 3.7399999999... and 3.74000000000... of the float 3.74
The first one uses an arbitrary number of trailing 9's and an ellipsis indicating that there is no end to the float string (in direct rejection of the finite declaration of the value 3.74
The second with unnecessary trailing zeros only indicates a bloated boneheaded alternate which deserves to be summarily prohibited. After all we have a nice rock-solid value 3.74 and alternate forms are adding an unnecessary cognitive load. The ellipsis character stream (…) after trailing zeros is redundant.

I feel compelled to keep repeating the obvious.
Even though in various articles on the Internet and chapters in textbooks promote such ambiguities we must reject them as the alternative would be the disintegration of our counting efforts. In this case we could be boneheaded and append an arbitrarily large number of the digit 9's which would stall counting even a single alternate form of any float.

7. It is the strict enforcement of this rule which excludes scientific notation - an alternate representation form for each float. 3.0 E-1 or 3.0 x 10^-1 both resolve to 0.3. Scientific notation is an algorithm which converts a mantissa and an exponent to a proper float

8. Floats can designate only floating point quantities and never integers, so 5.0 is invalid.

9. Additional modifiers such as the negative sign represent a floating point number and an algorithmic instruction (which is : "multiply this number by -1 or change the sign of this number, or use this number to subtract it from another number)

10. Implied positional meaning, for example 5.1^3 which resolves to 5.1 * 5.1 * 5.1

 Although we tried mighty hard to scrub the floats free of any algorithms, and manipulate them only as strings, that turns out to be an impossible and ultimately an undesirable goal. Floats will retain an inherent numerical base designator (2, 10 or 11 for example for binary, decimal and base-11) and will be called upon to use the base integer value when computing the so called "fake integer" values. Without such utility the algorithms which match floats to integers and vice versa, would be impossible.

Disclaimer : I do not advocate a draconian enforcement of these syntax rules in everyday commerce or for casual use. Violating these rules for the sake of brevity or for other reasons such as aligning columns of numbers for better readability makes practical sense.

But when we do count the floats our rules must stand both for the integer counting numbers and the floats. No exceptions. Without enforcing them it is not possible to do accurate and unambiguous counting. Note : In common use the float strings "1.0", "0.", "0.0" , " .0" or "1." may be accepted as integers. We however classify them as "sloppy integers" and as such are not valid.

Floats and **float_stories** are unfortunate auxiliary notations which grew out of algorithmic implementation constraints. I was simply unable to manipulate the computer software implementation of **floating point number** types with sufficient ease and competence, so I bailed and used the available **string** software type to define **float_story** variables. Such variables will be delimited by double quotes, e.g. "3.237"

Eventually **float_story** was abbreviated to **float** and as the current state of this manuscript will show. I unfortunately switch back and forth. (Yes, I will have to fix it in the next edition)

4.1 List of syntax rules, for decimal integer counters :

1.Only the 10 symbols [0,1,2,3,4,5,6,7,8,9] may be used. They are also called integer digits.
2. Leading zero allowed only for the number 0 (it is not a counting integer but an anchor, an indispensible placeholder and a powers of 10 multiplier)
Rule #1 excludes the popular alternate (sloppy) integer notation of **integer.0**

4.2 List of syntax rules, for floats being counted :

The rules for floats are more detailed and are described below :

The general structure of the floats is : **valid_integer.non_zero_integer_in_reverse**

This is the most concise definition possible. The next 8 rules simply elaborate
Note : An integer_in_reverse allows leading zeros but prohibits trailing zeros

1. Only the 10 symbols [0,1,2,3,4,5,6,7,8,9] and the decimal point ["**.**"] may be used
2. Must use at least two of the 9 digits [0,1,2,3,4,5,6,7,8,9] in any mix.
3. At least one of the digits used must be other than the digit 0
4. Use the decimal point ["**.**"] exactly once.
5. No trailing or leading decimal points are permitted.
6. Trailing zeros are prohibited.
7. A single leading zero is allowed if it is immediately followed by a decimal point.
8. There must be exactly one float string corresponding to a given float value.

Rules #2 and #4 mean that a float string is at least 3 characters long.

Rule #1 excludes + and – characters, scientific notation, e.g. 1.5 E-7 is prohibited.
(Note : Even the minus sign (–) is an algorithm, the algorithm of annihilation. Let 3 and -3 meet. They will resolve to 0 technically, but existentially there is nothing left. Zero itself is just that : nothing !

Rule #6 excludes 0.0 or 1.0 or any x.0 from being considered as floats. Although otherwise they appear to have the valid syntax, the prohibition is far more fundamental : they have integer counterparts of 0, 1 and x.

Rule #6 and #8 prohibits 5.1230 or trailing zeros, since it has the same value as 5.123

Rule #7 and #8 prohibits both 00.123 and 01.23: or multiple leading zeros since they both have their respective duplicates : 0.123 for 00.123 and 1.23 for 01.23

Rule #8 is a redundant declaration, it was included to drive home a very important point :
1. For each string of characters representing a float, there is only one possible value.
2. For a given float value, there is only one possible way to represent it in a string format.

5. Justification for rules :

What probably might be one of the leading causes of disagreement are the two rules which prohibit leading zeros for integers and both leading and trailing zeroes for floats. After all, what is the harm ?
 They all resolve to the same values, so we get identical results, are we nitpicking here about appearances.
Nothing could be further from the truth.
If we did allow such duplicates in our counting effort, we could fail in two major ways :

For example take the integer 1.   To count it using the lose syntax with leading zeros we would list 1, 01, 001,…. 0000000001

Then the floats :
1. We would never get past the first number 0.1, because then we would be allowed (nay, required) to count 0.10, 0.100, 0.1000 and so on. Here the multiple counting is obvious so we would catch it but what about the following scenario :   0.1, 00.100, …, 0000000.1000000

2. Now imagine that we would use 1, 01, 001, etc. to count floats 0.1, 00.100, …, 0000000.1000000
The entire counting project for the floats would collapse into a heap of irrelevant inconsistencies, with results we could never rely or replicate consistently.   If we allowed them then the floodgates would open to the ultimate boneheaded counting.

As I am fond to say : Open the door or of casual sloppiness an inch, there will be no shortage of boneheaded jokers who will drive a 747 through it.

I hope I conveyed a healthy respect for the Draconian (seemingly cosmetic) syntax rules…
They are the quintessential backbones of the entire counting project and the stalwarts of sanity.

Of course you might ask : "What is the right way to represent results when subtracting 3.3 from 3.3 or adding 0.3 + 0.7 ?   The answer is, use whatever syntax fits your application.
The rule matrix we presented here apply only to the "counting of the floats" project. Do your own thing.

In closing : without the rules above the floodgates will open and there would be unmanageable chaos.
It would be possible to generate an infinite number of unique strings mapping to the same values.

The various complete tree, level and bracket structures which we will employ later would get corrupted beyond repair. To avoid such outcome the counting algorithms must anticipate then prohibit sloppy, ambiguous syntax, which must be done before any processing starts.

That is exactly the syntax rules we presented do already.

As an added bonus, syntax rules promote and demand mental discipline but at the same time aid in forming logical and unambiguous mental constructs which are essential to come up with the most efficient and elegant conversion algorithms in the history of counting the floats.

All the syntax rules are coded and enforced by a single Maple-2018 procedure.

Examples of the Float Strings Syntax Rules © 2018 TMV			
Floating strings	Integer counting numbers	Validity	Reason
".".	N/A	NO	All 5 examples show floats with less than the required minimum length of 3.
"0."	N/A	NO	
".0"	N/A	NO	
"1."	N/A	NO	
".1"	N/A	NO	
"0.0"	N/A	NO	Trailing decimal 0, missing non-zero digit
"0.1"	1	YES !!!!	Proper syntax, the first valid float.
"1.0"	N/A	NO	Trailing decimal 0
"03.1"	N/A	NO	Leading zero must be followed by the decimal point
"0.10"	N/A	NO	Trailing decimal 0
".0.1"	N/A	NO	Too many & leading decimal points
"0.1."	N/A	NO	Too many & trailing decimal points
"0.1.1"	N/A	NO	Too many decimal points
"0.1"	1	YES	
"0.2"	2	YES	
"0.9"	9	YES	
"0.10"	N/A	NO !!!	Trailing decimal 0 !!!!
"0.11"	100	YES	Valid but out of sequence, has 4 characters instead of 3
"1.1"	10	YES	This is the correct 3-character, next-in-sequence float after "0.9"
"9.9"	90	YES	The last "3-character" float string
"0.01"	91	YES	The first "4-character" float string
"1.3E-4"	N/A	NO !!!	It is a float in scientific notation
"001"	N/A	NO !!!	This is not a float (and not even a valid integer

For future reference a compact version of the rules are listed below :

1. Only the 11 symbols [. , 0, 1, 2, 3, 4, 5, 6, 7, 8, 9] may be used.
2. Use the decimal point [.] exactly once.
3. No trailing or leading decimal points are permitted.
4. Trailing zeros are prohibited.
5. A single leading zero is allowed only if it is followed by a decimal point

6. What are fake integers ?

Fake integers is a term I coined for the duration of this project. I have found references to it here:
https://mathoverflow.net/questions/283172/fake-integers-for-which-the-riemann-hypothesis-fails
but the meaning is entirely different.

Fake integers are derived from any float by stripping it from its decimal point first, then removing all superfluous leading zeros if any. After the surgery the resulting string will be interpreted as a valid positive integer. Thus the negative powers-of-10 associations of the float will be converted to a zero or a positive powers-of-10 association.

Examples : Take the float 13.72, its fake integer equivalent will be 1372. For 0.1 or 0.01 or 0.00001 the fake integer value will acquire the integer value of 1 for each of the 3 floats listed.

An example on how to obtain the base float and the fake base integer from a known float string :

Float_story	"34.075"
1. Remove the dp and the 2 "-s to obtain the fake_story (decimal)	34075
2. Create the float_story template using matching x-s	xx.xxx
3. Fill up this template with all zeros	00.000
4. Change the rightmost 0 to a 1	00.001
5. If the dp is in the leftmost valid position (br_id = 1) then we are done	
6. Otherwise change the leftmost 0 to a 1 to obtain the float_story_base	10.001
7. Remove the dp to obtain the fake_story_base_decimal	10001

Since there cannot be reasonable discussion without counting and sequencing we will retain the implied non-negative powers of 10 for the decimal counting project.

For future reference let's list several application extensions associated with the decimal floats which are defined by the quintessential 11-member symbol table [. ,0,1,2,3,4,5,6,7,8,9]

The implied decreasing power of 10 attached to both integer and fractional parts.

The well known geometrical association of distances with the fractional number line.

The ubiquitous decimal point which - when inserted into any digit string - will impart the power of $10^0 = 1$ multiplier to the digit to its left and a power of 10^{-1} or 0.1 to the digits on its right.

The syntax rules which we listed above. [Ch 3b]

The digits from 0 to 9 indicate increasing numerical quantities which can only be described with a pictorial association with physical objects, in this case we do it with circles or "beads" as we frequently will refer to them : [0 >> o, 1 >> oo, 2 >> ooo, ... , 9 >> ooooooooo]

As much as possible, we will avoid any geometrical association involving the number line, or physical interpretations of implied powers of 10 numerical quantities. Such associations essential as they might be in the practical realm – are not relevant to the float counting enterprise.

Note that we are not restricted to the decimal or base-10 representation. Two other well-known and frequently used bases are the base-2 or binary with [. ,0,1] as the symbol table and the base-16 or hexadecimal variant using [. ,0,1,...9, A,...,F] for its symbol table.

Now, if the underlying implied extensions listed above are removed we would have the liberty of picking distinct or even repeating symbols which can be an arbitrary set of pictorial representations, for example : [A,$, %, m, n, x,x,q, ?,x @].

True, but if we did that we would have nuked our good old floating point number system. So for now we will stick with the base-10 decimal point floats for all discussions during the development of the counting algorithms.

Conclusion of this chapter :

What is left from this somewhat strange discussion is this :
We defined the following 2 classes of numbers :

positive integers : any string which uses one or more of the symbols : 0 1 2 3 4 5 6 7 8 9

floats : any string which contain exactly two sets of integers, separated by a mandatory symbol [.]

To facilitate coherent discussions we will use the common sense naming :

integer_part of real numbers : any integer which is situated on the left side of the separator

fractional_part of real numbers : any integer which is situated on the left side of the separator

We quickly note it here that based on our very strict syntax rules there are a large variety of float look-alike strings will not be considered real numbers, popular usage notwithstanding. I am not aiming to be a syntax police, however we are soon going to start to generate all the floats numbers in existence (at least in theory) there are employing a methodical, consistent and absolutely focused algorithm.

There cannot be ambiguities whatsoever or the show will not go on.

This is not nitpicking. Allowing those strings would result in unmanageable ambiguities as the same quantities could be represented more than one way. Sorry we will not allow phrases like "integer is a kind of real number or a real number could be a kind of integer" Such discussions lead to the nonsensical arguments such as : "an integer is a real number which is missing another integer and the separating character."

If the integer would be like a real number we would have defined it as a real number. The reason we have very precise definitions because they help us zeroing in on the essence of an entity and allowing us to design unambiguous algorithms to manipulate them at ease and without fail.

After all one of the claims I am about to make is that there are easily generated sequences of **unique floating point and** integer pairs. One very special mapping and the method of generation will be offered as a proof that there is actually one-to-one correspondence between bona-fide real numbers and the set of integers.

Chapter 4a

Counting agents, Countability, Enumeration
Boneheaded Counting (BHC)

One of the most powerful and highly demonstrative counterargument to various boneheaded argument is the unlimited number of independent integer sequences each of which we denote with a positive integer S. Within each sequence we generate integers according to the formula $N(S) = S * K + M$

S = 1	K=0	K=1	K=2	K=3	K $\to \infty$
M = 0	N=1*0+0 = 0	N=1*1+0 = 1	N=1*2+0 = 2	N=1*3+0 = 3	N=K

For S = 1 we can proceed only row-wise using a single agent, in essence we are doing the very well know positive integer counting by one. (0,1,2,3)

S = 2	K=0	K=1	K=2	K=3	K $\to \infty$
M = 0	N=2*0+0 = 0	N=2*1+0 = 2	N=2*2+0 = 4	N=2*3+0 = 6	N=2*K
M = 1	N=2*0+1 = 1	N=2*1+1 = 3	N=2*2+1 = 5	N=2*3+1 = 7	N=2*K+1

For S = 2 if we proceed row-wise we will need 2 agents to do the advancing one doing the even N=2*K numbers (0,2,4,6) the other the odd N=2*K+1 numbers (1,3,5,7)

S = 3	K=0	K=1	K=2	K=3	K $\to \infty$
M = 0	N=3*0+0 = 0	N=3*1+0 = 3	N=3*2+0 = 6	N=3*3+0 = 9	N=3*K
M = 1	N=3*0+1 = 1	N=3*1+1 = 4	N=3*2+1 = 7	N=3*3+1 = 10	N=3*K+1
M = 2	N=3*0+2 = 2	N=3*1+2 = 5	N=3*2+2 = 8	N=3*3+2 = 11	N=3*K+2

For S = 3 if we proceed row-wise we will need 3 agents to do the advancing " for N=3*K, N=3*K+1 and N=3*K+2 doing (0,3,6,9), (1,4,7,10) and (2,5,8,11) respectively.

Obviously the column-down-first, then row-to-right-next counting order will allow us to use only a single counting agent even for cases #2 and #3.
This agent then will proceed (0,1,2,3,4,5,6,7) for S=2 and (0,1,2,3,4,5,6,7,8,9,10,11) for S= 3.

In each scenario the counting will be accomplished in an identical manner. This will be the case when any finite N is compared with the multi-agent counting scenario.

The boneheaded counting method which is attempted by using an arbitrary number of sequences will result in an endless number of integers within the endless number of sequences.
Those who still do arithmetic with the infinity symbol, might attempt to informally denote the grand total as in the neighborhood or dimension of ∞ times ∞ (or ∞ to the second power).
But there must be a contradiction and no it is not the result of multiple enumeration of any integer.
These sequences will for sure account for each and every integer and do so exactly once.
The problem and the weird mismatch simply the result of attempting to do rational arithmetic with infinity, which is not possible.

1. Making the distinction between enumeration, counting and countability

When discussing countability the first roadblock we must clear is linguistic in nature.
Turns out that the verb "count" stands for two dissimilar actions.

Paraphrasing the bare-bone dictionary definitions for the verb **"count"** :
1. recite numbers in ascending order, up to the specified number
2. determine the total number of a collection of items

So back to the dictionary, there is another promising word : **"enumerate"**
1. mention a number of things one by one
2. establish the number of

Now we are basically in a corner as two distinct words in essence define the same thing.
Since there is no way to have an intelligent treatment of the subject if we use an identical word for two related but fundamentally different actions we will make an executive decision
and use two different words. The extended definitions applied to the treatment of the countability of floats are below :

1a. Count (always a verb) : Establish a unique one-to-one correspondence between consecutive positive integers and a collection of items. For the duration of this project the main focus of counting will be the positive floats. Example : 1 –> 0.1, 2 -> 0.01, 3 -> 0.001, and so on.
We usually count by starting from 1 then increment the count by 1. This is not mandatory but will serve our purpose. Counting is basically an act of annunciation of consecutive positive integers while pointing to an item … (potentially an endless succession of **screams !**)

1b. A collection of integer or float items which we can count have the property of **countability.**

1c. Enumerate (a verb) : Obtain the number of items in a finite set by **counting them first**,
then subtracting the initial sequence number from the final sequence number and adding one.
In the previous examples the number of items equals : 3-1+1 = 3

Enumeration is counting then evaluating the total number of items using the formula :
$$total = max\_count - min\_count + 1$$

1d. Items we cannot enumerate : If the number of items is unknowable or it is changing while the act of counting is performed **we cannot accurately enumerate** those items.
We have coined no special name for such collection of items, but we insist that they not be called uncountable.

2. Countability properties of positive integers :

A sequence of positive integers starting from 1, incremented by 1 and having no upper bound **cannot be enumerated**. They are however **countable**.

A sequence of positive integers starting from 1 incremented by 1 and ending at VBN **may be enumerated** and of course as such are countable.

3. Countability properties of positive floats :

A sequence of positive floats starting from any value, incremented by using any rule and having no upper bound **cannot be enumerated**. They are however countable. Please remember that counting is simply an act of acknowledging a particular float by announcing the next integer positive integer counting agent.

A sequence of positive floats between a start and finish value still **cannot be enumerated unambiguously**. They are however **countable**, but such action if undertaken without foresight it will not terminate and the whole process will make little sense..
Example : start_float = 0.1 finish_float = 0.2. We can start counting as follows :
0.1, 0.01, 0.001,...0.0001 and so on. We obviously get derailed into a monotonously decrementing so called boneheaded sequence from which there is no recovery. There are an undetermined number of such dead-end sequences, giving us ample opportunity to mess up.

4. How to count between any pair of floats.
A universal, fool-proof, minimal distance way to count between a start and end floats
was the mission and backbone of this project. We managed to design and implement such "perfect" counting algorithms. Without exaggeration the end result is an incredibly efficient, limitless and truly elegant pair of algorithms. With the help of these algorithms we will be able to debunk any and all statements which claim that the floats between two limits cannot be enumerated.

5. Fair versus boneheaded counting and how to avoid the latter

Fair counting is a method which will provably count all of the numbers up to a certain
maximum number without limits. For endless sequences that is the best we can ever do.

For example : The integer counting of 0, 1, 2, 3,...9, 10,...100 first counts all numbers with the representation size (RS) of 1, then all of them with RS = 2 and so on. Although there are an unlimited number of integers (which means we can never finish the counting) we can show that we can count all numbers up to and including a given size and that there is no roadblock in sight. As a counterexample, counting only the even numbers would definitely leave half of all integers of every size uncounted. Thus such exercise earns the BHC title

What is critical, it is not the proof that the counting can be eventually finished.
If it was we would have to classify integers as uncountable as we can certainly
do not have the last integer in sight.
Maybe it is the ability to prove that the counting does not dead-end or funnel into one specific sequence from which there is no return. "No return" means that other than the members of a specific number sequence, no other numbers will ever be counted.
For example 0.3, 0.33, 0.333 is a sequence where we only include floats which exclusively use the digit 3. This obviously excludes a majority of floats.

6. Fact and fallacies of boneheaded counting :

1. Sporadic boneheaded counting of floats unfairly classifies the set of floats as uncountable.
2. That there is no dispute that integer sequence numbers are countable.
3. Most float-specific boneheaded counting has its counterpart or equivalent boneheaded counting in the integer sequence number domain.
4. Most of the float-specific sequences can be disqualified on their own demerit while for others we can use the equivalent integer counterpart's uncountability as proof that the float sequence is not fair either.

7. Examples of boneheaded counting :

Example #1 : Counting only the even integers (we will never count the odd integers)
Example #2 : Counting only the positive integers (we never get to count the negative numbers)
Example #3 : Count only the powers of 10 integers, e.g. 10^0, 10^1, 10^2
Example #4 : Factorials, which is a good example of an algorithmic dead-end counting
Example #5 : The claim is that since 1/3 evaluates to 0.3, then 0.33, then 0.3333 and so on without end, it proves that the floats are uncountable, since all counting efforts are dissipated trying to follow a single thread.

Let's hold there for a second and note that if we remove the period and the decimal point then we end up counting the following integer sequence 3, 33, 333, ..., 333333... Applying the same reasoning as was done for the 0.3, 0.33, 0.333... sequence we would have to render integers uncountable. Both are boneheaded techniques which we reject summarily. (at least for the purpose of proving non-countability of either sets)

Example #6 : 0.1, 0.01, 0.01... strip leading [0.] then reverse the digits to obtain the integer series :
1, 10, 100, 1000 ...10^n this is the same logic and same conclusion : either both floats and integers are countable or neither of them are.

Example #7 : There are more floats between 0 and 1 than there are integers.
True or False ?

Example # 8: Boneheaded counting caused by permissive syntax definition of floats:
These are easy to demonstrate as the pattern is probably a unique combination of simplicity and hopelessness.
Take the single integer whose value is 1. We can immediately render this integer uncountable all by itself if we allow leading zeros. The sequence of integers
1, 01, 001, ..., 000000000000001 is uncountable, yet they share the same value of ONE !

The float counterpart of 0.1 is even more pathetic as we can create an inconsequential monster going in two direction : We may allow both leading and trailing zeros.
The sequence of illegal floats 0.1, 00.10, 000.100, ..., 00000000.1000000 is uncountable, yet they share the same value of 0.1

Exercises in boneheaded counting should make us aware that careless, unfounded claims about countability will gain traction if we allow it. We must start out with absolutely bare-bone minimalistic (and non-apologetic) syntax definition for both the integers and floats.
Otherwise we will get lost in the labyrinth of irrelevant and completely wrong competing claims.

8. Argument for and against bonehcadcd counting:

8.a For : Generate floats by putting the entire integer-like sequence on the right side of the decimal point (dp)

Floats	0.1	0.2	0.3	0.4	...	0.9999
Counters	1	2	3	4	...	9999

Floats	1.1	1.2	1.3	1.4	1.5	...
Counters	None left, floats with whole parts $\lessgtr 0$ remain uncounted					

Floats	0.01	0.02	0.001	0.0001	0.0999	...
Counters	None left, floats with whole parts = 0 and with digit zeros after the decimal point remain uncounted					

Claim : All the integer counters are used up by a single sequence of floats whose whole parts (on the left of dp) equal to 0 and whose fractional part does not have a leading zero. (Note : we admitted floats with trailing zeros e.g. 0.10)

Observation : Based on the sample pairing most floats between 0 and 1 remain uncounted.
Conclusion #1: There are more floats between 0 and 1 than there are positive integers.

8.b Against : Use a different association to prove conclusion #1 False

1. Remove the "0." prefix from each float
2. Remove all trailing zeros as well
3. Reverse the order of digits of the truncated digit sequence.
The resulting digits will constitute valid integers exhibiting one-to-one correspondence with the original floats.

Floats	0.1	0.01	0.0001	0.9	0.19	0.123	0.00010007
Counters	1	10	1000	9	91	321	70001000

Footnote #1 : Building chimneys the boneheaded way

There is a very strange history of bringing one example e.g. repeating decimals to prove uncountability. How come nobody ever said : Hey, why don't we count a few digits of 1/3 then a few digits of π, then that of sqrt(2). If they did that, from there it would have been an easy logical conclusion to take all possible reals of a given size and count those, before moving on to a longer one (higher character count)

What is missing in each pathological case is proportionality, trade-offs and common sense.

The two extreme pathologies of counting have an interesting counterpart in human affairs. Consider a simple task of building chimneys for the dwellings of a small settlement.

Something goes wrong and two polarized camps fight over what is the right way to proceed.
One advocates the building of a single, as tall as possible chimney using up all available bricks.

Construction starts, the chimney becomes a veritable Taj Mahal, powering one furnace in the middle of the village. The few people who gather around it survive for a while but the rest of the village freezes to death. (this is the analogy for the 00000000.1000000 float which uses up all the integer counting agents.)

So they swing the other direction. Arguing for more egalitarian distribution of bricks and human effort, they decide to hand everybody a single brick in the village. Next day they hand out the second brick and so on. Turns out that when all the bricks are gone, every furnace is only 50 percent complete. Once again disaster is the price to pay for boneheaded distribution.
(this is the human tale of the 0.3, 0.33, 0.333... sequence).

The obvious solution is to distribute the bricks to a number of houses with each receiving enough bricks to complete one furnace sufficient to heat the house. Eventually the bricks will be gone before each family gets to build their furnace. But enough will be operating so that the villagers can share houses for the duration of the winter. Everybody survives and will have 3 seasons to mine enough clay to build the rest of the furnaces before the next winter comes.

Footnote #2 : What if integers are also uncountable ?

Although it is well-removed from the float countability/cardinality debate, this would be a good time to mention the quick back-and-forth argument which goes like this :

Proponents of uncountability would present a sequence of floats 0.3, 0.33, ..., 0.333333...claiming that it takes up all counting resources so the entire collection of floats is sure uncountable.

Proponents of countability would respond with the integer counterpart of that sequence :
3, 33, 333, ..., 333333... claiming that for purposes of deciding countability it is just as good as the float sequence.

Then they would drive home the point : "But we know that all positive integers are countable, which means that the 3, 33, 333, ..., 333333...sequence cannot be an indicator of the absence of countability : not for integers and by analogy the 0.3, 0.33, ..., 0.333333...cannot be used for floats.

That sequence of reasoning is definitely sound, but is missing a delightfully simple proof.
Here goes : The **boneheaded counting of 3, 33, 333, ..., 333333** proceeds by assigning counting agents to each member of this sequence as shown in the table below. We can see that the sequence "unfairly commandeers the integer counters from the start, never allowing anything else to be counted.

3	33	333	3333	33333	....
1	2	3	4	5	.....

However, the **fair and equitable counting of that sequence** would tag with sequence numbers as shown on the second table below. Each and every number in the sequence would have to wait its fair turn to be assigned a sequence number 33, 333, ...33333... (which in the case of integers is trivial but in the case of floats is not) The other numbers would be utilized to count every integer which is not part of the sequence.

1	...	33	...	332	333	...	33332	33333	....
1	...	33	...	332	333	...	33332	33333	....

If this sounds painfully trivial too you it is. Which brings up a good point : Why did it take over 125 years to act on it by somebody ?

Footnote #3 : Can't count Powerball actual winning numbers
Let's demonstrate uncountability with an everyday example, the Powerball Lottery drawing. Every week the jackpot combination is assembled by pulling 5 out of 69 then 1 out of 29 sequentially numbered balls.

That act of selecting the 6 numbers is actually an algorithm resolving to a set of 6 integers.

There are **292,201,338** possible Powerball number combinations. As big as this number is, it is obviously **possible to generate all combinations then enumerate them once and for all.**

Now consider that since the inauguration of the Powerball game there have been only about1500 actual winning jackpot numbers. It is trivial to count and enumerate them up to the present. Yet, the theoretical set of actual winning numbers are hopelessly uncountable as they are an open and undefined set.

Imagine for sake of simplicity that by sheer accident every week the numbers 1, 2, 3, 4, 5 and 6 will be pulled. That can go on indefinitely.

We can **count** the weekly drawing of each jackpot but we will never be able to **enumerate** them.

Footnote #4 : Some boneheaded sequences are worse than others

The commonplace boneheaded sequence is when we drive into a tunnel never to
resurface again, for example : 0.3, 0.33, …0.333 …
We ignore all other sequences and individual floats but at least we count different floats.

Now let's do trailing zero boneheaded counting : 34.2, 34.20, 34.200, 00034.200000 and so on.

We would be doing no meaningful counting at all as we would be stuck on the value of 34.2 forever. Total waste of CPU cycles (and our time…I am wasting it now by trying to explain the obvious)

Chapter 4b

Value and String Length Divergence for Floats

There is no such a thing as a "smallest float", …only a "smallest float within a given character length !

1. Disprove false claims of uncountability caused by boneheaded counting

An example false claim: "Even simple sequences of floats like 0.3, 0.33, 0.333… and 0.1, 0.01, 0.001, … make the floats uncountable. By diving into any of these sequences the rest of the floats are ignored and remain uncounted. "

Response : Such unfair or boneheaded action does not render the floats uncountable, it only exposes the errors committed by human counting practices. If such erroneous counting processes were applied to integers it would seemingly render them uncountable too.

Examples :
  Integers which only have one kind of digit 3, 33, 333…
  Counting only even integers [2, 4, 6,…]
  Counting only positive powers of 2 [2, 4, 8, 16,…]

2. Implement fair and proportional participation by floats in the counting process

2.1 Use integer counting agents

Taken a finite set of objects then we can enumerate them by using the positive decimal integers as counting agents. The **"no leading zero"** design has made the positive decimal integers the #1 bead counter or counting agent of the civilized world as it removed 00, 000, 010 etc from the list of counting agents. The worst feature of leading zeros for counting agents would generate "leading-zero" strings.

Such duplicates might have been differentiated by the virtue of the number of leading zeros and not by their values, which would resolve to duplicates.
But the rest of the counting agents would be differentiated by inherent values.
Such two-tiered value resolution would create mass confusion and thus had to be abandoned.

So when observing complete decimal integer trees we might have noticed that the digit 0 is missing on the first level. Because our counting agent is value-oriented as opposed to being driven by the geometry of the Complete Tree, we must start the counting with 1 .
Otherwise the value of the counting agent would be trailing behind the counted objects by one.
Only when we announce "one" do we move the first bead.

However, if we counted using the Complete Tree using an arbitrary alphabet without implied values, then we would not match values but pictures of strings which reside in the nodes of the Complete Tree.
So for an alphabet [A, B, C] we would match "A" with one, "B" with two and C with 3 beads. AA would correspond to 4 and AAA to 13 beads.

2.2 Count with piles of objects and beads

Counting with integers is an act which has a physical equivalent. Assume two piles O and B, where O is arbitrary collection of objects which need to be counted and B contains a number of equal-size beads, to be used as counting agents. Between the two piles there is a counting agent, either a human being or a robot. The agent simply takes one object from pile O and puts it to his left, into pile OC then one bead from B and collects them in another pile BC on his right.

When the question "what is the final count" asked, nothing need-to-be (as a matter of fact nothing can be) said.

Pointing to the newly formed pile BC (containing the beads which were moved) is the answer. This was simple enough until something else needed counting. Then the pile of beads had to be merged with the original master pile to perform the next counting task. No record was left of the prior counting.

3.1 Symbolic counting of arbitrary object sizes

That is why we invented ultra-sophisticated symbolic counting usually performed instantaneously by machines instead of us.
Piles of beads were made obsolete and record keeping moved from chiseling notches onto cave walls to memory cards which can hold information far exceeding the storage capacity of all the caves on our planet.

Eventually with the invention of the floating point numbers (or floats) the counting has matured from being able to handle only equal-sized beads to counting objects of arbitrarily small sizes. But oddly enough, somewhere along the way we lost our nerve and have decided that floats themselves are not countable. As I will show that this conclusion which now survived for over 126 years is utterly wrong.

Floats are eminently countable !!!

3.2 Limits of counting

One can raise a theoretical objection to our method of counting that no matter what, it is still an open-ended never-completed task. Well, in theory yes, but not in a practical sense. We invented integer numbers so that we can measure the extent of giant galaxies, but created floating point arithmetic to size up a mosquito, a bacteria, virus an atom or an electron. They are all physical objects in our world. The precision and magnitudes we hope to handle can be represented by up to 100 digits after the decimal point.

To go the other way to the universe we can observe, for now we are all set if we use 100 digits to left of the decimal point. That is a total of 200 or so floating point digits (the number of digits is arbitrary we are trying to make a point here).

Then any numerical value associated with an object or used to describe an act in the physical world can be accommodated with plenty to spare. Although the numerical output of an open-ended set of algorithms is undefinable and worse – unlimited, that does not prevent us to count and measure everything in the world around and within us.

3.c One-to-one counting

However, we should not attempt to count each and every float ever existed because that task cannot be accomplished.
Instead we have invented and implemented the absolute best, counting **on-demand, one-to-one association.** It is a reciprocal relationship which has existed in the universal library since the dawn of time or way before that. We simply authored two Maple-2018 procedures to gain access to it.
(The reference to an ephemeral universal library is not entirely facetious, more on that later)

Here is what we have : For any legal syntax float (call it : float_story) I have provided the means to compute the corresponding integer sequence number (call it : int_seqno) which can be recalled by invoking a Maple-2018 procedure of the form :

$$\text{int\_seqno:} = \text{find\_int\_seqno\_from\_float\_story (float\_story)}$$

This int_seqno cannot be destroyed, altered or modified no matter what. No float_story will ever lose its one and only one designated counting number. Think of it as a sort of eternal ID. It saddens me that we humans, who are a heck of a lot more significant and capable than a lowly float_story, do not have an eternal int_seqno assigned to us. Once we kick the bucket, our only ID the Social Security Number will either be retired or assigned to the next contestant coming into this world.

3.d Which is better, decimal or binary base ?

The decimal integer vs. floats counting algorithm was the first I completed. Counting and devising algorithms in decimal is far more intuitive and familiar than doing it in binary.

However, binary is the quintessential number system, the root of all others. It is the first and only counting and algorithmic base we need. There is no way to have anything to do with numerical algorithms if only a single number, zero is used. We must have a first waypoint, unity or 1, a goalpost on our way to the unimaginable heights we took mathematics or mathematics took us. Beyond that no other digit is needed, except for balancing utility with convenience. Thus binary is fundamental, but basing practical mathematics on it would condemn us to unnecessary drudgery, so we use decimal or base 10 as a perfect trade-off between stark utility, happy productivity and leverage. We use decimal in everyday dealings, but we call for binary when we design machines or when we need to demonstrate the ultimate elegance and simplicity of some algorithms. Since I implemented all-base, general counting algorithms, binary and decimal bases are both options.

4. Counting integers is possible either by value or by representation size
Countability for open ended sets, specifically for positive integers and floats does not
hinge on whether it is possible or not to account for every item in the set in a finite effort.
Neither the integers, nor the floats can ever be **enumerated** unless some specific criteria is available.
Let's take the positive integers : we can limit the task by two ways :

4.1 Count integers by representation size

We specify an upper limit to the number of characters used to specify the integers.
For example count in lexicographic order :

1. all the 1-digit integers from "0" to "9" (we have 10),
2. all the 2 digit integers from "10" to "99" (we have 90)
3. all the integers 3-digit integers from "100" to "999" (we have 900)

4.2 Count integers by inherent value size

We specify an upper limit to the underlying value of the integers.
For example count by numerical values of the integers:
1. all the integers from 0 to 9 (we have 10),
2. all the integers from 10 to 99 (we have 90)
3. all the integers from 100 to 999 (we have 900)
As we can see, for positive integers the "by-character" and "by-value"
counting methods are identical, we can go either way.

Let's digress here for a second : It is obvious to almost everybody that the
positive integers are eminently countable. But what about all the integers which include the
negative side ?
If not careful we can immediately fall for the boneheaded sequence which is to count the positive
integers first, then go after the negative ones. Wrong, as doing so we would never finish counting
the positive side. The solution is simple : count by switching signs at some arbitrary waypoints :
Example : 0, 1, -1, 2, -2 and so on (or 0, 1-> 9, -1 -> -9) and so on

Note that we will remain focused on the simplest two sets : the positive integers and positive floats.
The counting principles will be sufficiently demonstrated using those two;
No need to beat the subject to death.

4.3 Failed attempt to count floats by inherent value size first

For sake of simplicity let us confine ourselves to positive floats less than 1.
That is, the integer part of the float is set to 0. Before we start counting,
let us find the upper and lower values of the floats

And we are TOAST !!!!!!! There is neither UPPER nor LOWER value limits to
the fractional part of a float unless its representation size is constrained (fixed)
The sequence starting with 0.1 will progress to 0.01, then 0.00000…..0001 approaching zero, but
never reaching it, while at the other extreme starting with 0.9 will progress to 0.99, then 0.999…..999
approaching one but never reaching it either.

4.4 Floats can be counted only by representation size first, by value size second

We specify an upper limit to the number of characters used to specify the floats.

For example count in lexicographic order
1. all the 3-character floats from "0.1" to "9.9"   (there are 90 of them)
2. all the 4-character floats from "0.01" to "99.9"   (there are 1710 of them)
3. all the 5-character floats from "0.001" to "999.9" (there are 25200 of them)

What is the smallest valued float? If you are familiar with the relevant literature the stock answer will
be : we do not know, as the floats may represent arbitrarily small and ever decreasing quantities. True.
Each float has an associated implied quantity, which cannot be explained any further using floats, it has
to be connected to physical quantities.

For example let's take a stick of some length, and declare that its length is of unity or the integer number
1. Then if we cut the stick in half, we can represent each of the pieces as having the length of 0.5.
Subsequent cutting will result in smaller and smaller pieces whose lengths will be 0.25, 0.125, 0.00625
and so on. Now, you might have noticed an interesting contradiction. While the sticks we represent are

getting smaller, the character representation of their lengths will grow. Wow! As the sticks are approaching, then passing atomic sizes, the representation sizes will grow past the material resources of the universe assuming that each digit requires a non-zero number of atoms to represent.

Since floats also have integer parts, as we increment that side of them, their representation also grows. So on the integer side of the floats we do have a one-to-one correspondence between implied value and representation size. This divergence of behavior appears to make counting the floats even more challenging. But it is only appearance and we will show shortly that enumerating the floats within representation brackets is eminently doable.

5. Various counting options

5.1 Base-10 integers with leading zeros allowed
The set of variations which can be obtained by traversing a size-11 Complete Tree with the alphabet of : [0, 1, 2, 3, 4, 5, 6, 7, 8, 9]

5.2 Proper decimal counting integers using a near-Complete Tree
The subset of variations which can be obtained by traversing a size-10 near-Complete Tree using the alphabet of : [0, 1, 2, 3, 4, 5, 6, 7, 8, 9]

On level 1, the set if available symbols excludes 0, that is the set of available nodes is restricted to [1, 2, 3, 4, 5, 6, 7, 8, 9]. The rest of the levels are identical to "case #1 "

5.3 The well known string-based decimal integer counting scheme
Instead of using Complete Trees we advance to the next integer by executing a cascading-carry operation : 1, 2, 3, 4, 5, 6, 7, 8, 9, 10(carry), 11,....99, 100(carry) and so on.
All computer algebraic applications use this "invisible" built in algorithm.

5.4 Base-11 integers with leading zeros allowed
The set of variations which can be obtained by traversing a size-11 Complete Tree with the alphabet of : [0, 1, 2, 3, 4, 5, 6, 7, 8, 9, A]

5.5 Decimal candidate floats
They are the set of variations which can be obtained by traversing a size-11 Complete Tree with the alphabet of : [. 0, 1, 2, 3, 4, 5, 6, 7, 8, 9] [see Chapter 5b]

5.6 Decimal valid floats
They are the subset of the decimal candidate floats, obtained by enforcing a consistent, common-sense syntax which are a stricter version of current floating point syntax rules.

6. Using Complete Trees to count floats
[see Chapter 5b for detailed treatment]

6.1 There is a 1-to-1 correspondence between any two Complete Trees
This is true provided that their alphabets are of the same size and each alphabet is a set of unique symbols without repetitions.

6.2 We can always count floats using a base-11 integer counting agent

6.3 Integer counting agents can always count integer counting objects
Holds true, regardless of the base of either. This is because integers of any two bases are one-to-one correspondence with each other.

6.4 Countability of floats is independent of the ratio of character sizes

The ratio of the character size of the counting agents to the character size of the counted numbers (floats, decimal or any-base integers) has no bearing on the countability of the counted numbers.

Footnote : How to Prove impossibility ? It is not enough to demonstrate one failure.

Instead it must be shown that all possible attempts fail. Of course we also must prove that we came up with a proven method of accounting for all possible attempts. (it is not enough to say we tried 1000 cases and they failed. In addition we have to show that (1) either that there are only 1000 cases, or to show that our proof a few cases covers all possible cases (proof by induction ???)

On the other side if we only show one possible working case then we disproved the contention of impossibility.

Complete Tree Structures Definition
One-to-one correspondence between base-11 integers and base-10 floats

Introduction

Up until now, nobody had any idea how to sequence the floats.
This has changed. There is an ordering arrangement of the floats which is so elegant and simple
that it can be mastered with ease without the aid of being associated with pre-determined
integer counters or Complete Trees.

1. Proving Countability of Floats Using Twin Complete Tree Structures

Up until this point we have managed to reject claims denying the countability of floats,
yet we never actually **proved that the floats are** countable.

That is true, so let's proceed. We will construct, then compare the contents of two 11-symbol
Complete Tree Structures (CTS) and demonstrate one-to-one compatibility of the node values.
This will prove without doubt the countability of base-10 floats using base-11 integer counting agents.

Positive integers are the wizards of counting, using them we are able to enumerate any collection
of objects by assigning a number starting from 1 and increasing this number by 1 without limit
(to get ready for the next assignment)

So self-counting is trivial for positive integers. 1 is the first in the order, followed by 2,3,...n.
That is, the sequence (or inventory) number of a positive integer is simply itself.

2. To count floats we can do :

1. Employ brute force, by generating complete trees then apply syntax to throw out those
generated strings which do not conform to predefined float syntax rues shown in **Chapter 3b**

2. Using suitable algorithms to access a very special virtual Level and Bracket pyramid (L&B)
data structure which stores a float_story and its corresponding int_seqno in the same location.
See Chapter 6a for the theory, structure and access methods of the Level and Bracket pyramid
and its associated algorithms.

3. Selecting the right data structure :

The well-known methods of wrestling with the traditional data structures with nodes, pointers,
memory management overhead, tree balancing and re-balancing are a difficult and messy proposition.
 I can think of only one redeeming feature :

They do work for limited data sets. Many times it is the only way to make an application feasible.
But the sizes of those structures are always bound by the available hardware, which of course makes
unlimited random access impossible.

However, there is another very unique and exceptionally versatile hybrid storage construct which
will help us immensely. Let me introduce the Complete Tree Structure (CTS). You will be
astounded as it is by far the simplest and most economical in the universe as it takes up zero
space by the virtue of being virtual (no, I could not resist that !)

For counting and two-way pattern assignments it is the quintessential egalitarian backbone of many algorithms as everybody can use it any time, simultaneously and without limits.

By comparing 2 complete trees of different bases we can create a binary then decimal or base 11 and record the 1:1 correspondence between the nodes and their contents.
This makes the CTS a very versatile and powerful counting tool.

The most famous and most frequently used variant of the Complete Tree is the Near-Complete Decimal Integer Tree. It is a mouthful, but by listing several of its contents everybody involved in counting will instantly recognize it : Here goes : 1, 2, 3, 4, 5, 6, 7, 8, 9, 10, 11, ...99, 100, 101,...,999, 1000, 1001,...,9999 and so on.

This is the CTS which produces the positive decimal integers so naturally that we can do it in our sleep (or before falling asleep as we count sheep) This CTS contains by its design the cascaded carry rules, which enables us to perform the 4 basic arithmetic operations : add, subtract, multiply and divide using pencil and paper. Why the "Near Complete restriction ? Because CTS structures for positive integers generate number strings with leading zeros resulting in duplicate values, for example : 0, 00, 01, 001, 0344 and so on.

Duplicate values introduce ambiguities which in turn force extra processing to eliminate them. Even worse if we insist on allowing duplicate representations for the same values then we would be carrying the burden of accommodate those duplicates which eventually would extra logic to be included to recognize them.
We can quickly dispose of its definition :

4. Defining the Complete Tree Structure or CTS :

CTS is a hierarchical storage arrangement containing an arbitrary number of nodes. Its mission is to store, then allow access to special data items which are always a sequence of predefined symbols. Let AS designate the number of symbols in the alphabet The set of valid symbols are predefined in a list we call an alphabet. There are no duplicates allowed in the alphabet, however there is no limit to how many times and in what order the symbols in the alphabet may be concatenated to form a story.

Now, as the picture shows we organize our symbols into levels using a very strict rule. On the first level we place one instance of the alphabet. On the second level we lay out AS copies of the alphabet, so we will have AS * AS symbols or AS^2 symbols. We proceed the same way for subsequent levels without limits.

It is easy to see that a level L will have AS^L nodes containing symbols. After summing the levels, the grand total of all nodes for the Complete Tree will be $AS*(AS^{Lmax} - 1)/(AS-1)$. With one single exception (which will always have only AS-1 nodes on the first level (to avoid leading zeros of the integer counting numbers), we will always deal with Complete Trees, meaning that there will be full house.

All possible nodes will be laid out and every one of them will be AVAILABLE. If your immediate reaction is that this will make for a very boring repertoire of strings don't worry.
The nodes will be AVAILABLE, but we will select only a specific subset of those to PERFORM.

Turns out that physically the **CTS** is the world's most boring tree since it can't even store a single tangible bit of information. Yet in spite of being virtual it can be tricked into doing magic. We will define a configuration which will allow us to pack the vast infinity of the floats into two reasonably sized algorithms acting on the **Float Complete Tree** or **FTS**. Even more impressively other than the miniscule memory requirements for the algorithms and housekeeping utilities, there will be no storage requirements.

For our counting enterprise we will predominantly use two alphabets : [0 -> 9] for positive integers and [. 0-9] for positive floats. (there are two exceptions : the binary float alphabet [. 0 -> 1] and the base-11 integer counting alphabet [0 – 9, A] (where A represents the integer value of 10)

The binary, decimal or any-base CTS is a true algorithmical wonder. In over 45 years of various software activities I have never seen or used anything which comes even close in simplicity, performance, robustness and wide range of applicability.
Its explanatory and concept proving-feature if not legendary today, will become that way soon enough.

Floats must be represented so that can be handled by various selection and display algorithms and equally important in a way that they make sense to us. This is accomplished by strictly defining the rules which will ensure that only legitimate floats are produced. Turns out that in the base-10 domain, the entire set of positive floats can be represented by the subset of the complete variations of 10 digits and the decimal point.

Note : Variations of a given symbol set are produced by obtaining all permutations of all possible combinations of the symbols comprising the set. However not every variation will correspond to a valid float.
For example 0000000 or .0..1...2..3 are valid variations but they are regarded as nonsensical floats.

5. We have reached several pivotal conclusions and solutions :

5.1 The Cantor Diagonal Argument (CDA) does not offer a relevant guide for the countability (or lack of it) of the floats.
Solution : Ignore the CDA and all of its derivative conclusions.

5.2 The reason floats were classified as uncountable was the universal practice of attempting to count the undefinable float outputs of an even more hopelessly undefinable set of algorithms.
Solution : Only count the floats themselves, never the algorithms.

5.3 We also concluded that a second reason for the perceived uncountability of floats was the careless and unjustified application of the so called **boneheaded or tunnel-vision** counting, where a particular endless sequence was followed to the exclusion of all other floats.
For example : 0.3, 0.33, 0.3333…333 or 0.1, 0.01, 0.0000….1 and so on
Solution : Design and implement fair round-robin counting sequences.

5.4 Compounding the false uncountability claims explained in 2a. and 2b. was the sloppy, ill-defined and haphazardly enforced floating point number syntax.
Solution : Define a minimal, very precise and at the same time provably sufficient rules of syntax.

5.5 The only way to count floats is to proceed from minimum toward maximum sizes.
Solution : Design and implement a "smallest toward larger size and increasing value within a given size" counting algorithms.

I will methodically enforce in this body of work all the solutions I outlined above while striving for algorithmic correctness, elegance, parsimony and simplicity. Let us proceed then.

6. Stages of Counting items :

There are various stages of counting a collection of items going from primitive to sophisticated stages.

Stage 1 : Counting using physical piles :

Assume that we have a number of items in pile A and plan to use a number of uniform-sized counting beads in pile B to count all the items in pile A. To accomplish this task we designate two auxiliary piles AC and BC.

We stand between the beads and the items to be counted. Then we simply move an item from pile A to pile AC and simultaneously move one bead from pile B to a bag BC. When all the items have been transferred we are done. The number of beads in bag BC is our record, representing the initial number of items in pile AC. If somebody inquires about the outcome of the counting we can't say nothing – we simply hand them the bag of beads.

Stage 2 : Associate a set of symbolic counters with beads.

Instead of using physical objects we define a permanent link between the number of beads and the counting agents. These symbolic counters are the well-known positive integer sequence numbers ranging from 0 to an unlimited number. Each symbol corresponds to a unique number of beads which are incremented by one, leaving no gaps.

	O	OO	OOO	OOO O	OOO OO	OOO OOO	OOO OOO O	OOO OOO OO	OOO OOO OOO
0	1	2	3	4	5	6	7	8	9

Stage 3 : Define a cascading-carry algorithm.

This algorithm employs the symbolic counters which will extend their range from 1 to an arbitrarily large quantity.

O	OO	OOO	OOO O	OOO OO	OOO OOO	OOO OOO O	OOO OOO OO	OOO OOO OOO	OOO OOO OOO O
1	2	3	4	5	6	7	8	9	10
OOO OOO OOO OO	OOO OOO OOO OOO	OOO OOO OOO OOO O	OOO OOO OOO OOO OO	OOO OOO OOO OOO OOO	OOO OOO OOO OOO OOO	OOO OOO OOO OOO OOO O	OOO OOO OOO OOO OOO OO	OOO OOO OOO OOO OOO OOO	OOO OOO OOO OOO OOO OOO OO
11	12	13	14	15	16	17	18	19	20

Note : The **cascading carry** may be performed by the well-known integer addition as
follows for decimal counters :
For each count we increment the last digit by 1 unless it is already 9.
When the last digit of the current count is 9 we set it to 0 and increment the next digit to the left.
We repeat these steps until all the digits of the current count are 9. Then we replace each 9 with 0
and prepend a digit 1. See the example below :

Current	0	1	8	9	12	19	99	8999	99999
add 1	1	1	1	1	1	1	1	1	1
Result	1	2	9	10	13	20	100	9000	100000

Stage 4 : Abandon the beads and use the set of symbolic counting agents exclusively.

From now on we will refer to the symbolic counting agents as positive integers or integers for short
as negative values are non-existent in our counting enterprise.

The **cascading-carry** is universally known and is self-evident when applied to decimals integers.

However the it becomes somewhat tricky when we use it for binary, base-11(floats) or hexadecimal
counting. It is especially tedious and error-prone process when the counting symbol-set contains an
arbitrary mix of numerical and alphabetic or other symbols.

Try counting the hexadecimal floating point numbers with the symbol set : [0..9, A..F, decimal point].
It is no longer trivial. Let's see if we can find something easier to handle.

Stage 4a : The answer is Complete Tree Structures (or CTS for short)

What is less known that the cascading-carry of counting has its equivalent virtual data structure CTS.
with the following properties :

1. It is virtual, defined only by its appropriate access algorithms. Its nodes are uniform and have
a multi-way connecting hierarchy.

2. A complete tree is constructed by linking a potentially unlimited number of nodes in an organized
sequence.

3. There is a well-defined relationship between the contents of each node and the location of the nodes.

4. The structure of links and the position of the nodes force the storage and access method of the CTS.

5. Without such a relationship the deposit and retrieval of data would require linear "on the ground"
   sweeps eventually chocking off access to all but the smallest possible CTS .

Now, a visual representation of a complete tree is in order. We will use a base-3 or ternary tree so the
details of the hierarchical structure and the associated digit strings stored can be demonstrated without
being hampered by the sheer number of nodes.

The complete tree structure can be scaled up to an arbitrary size and may be used with any application-
specific symbols set. The fundamental ternary digit list : [0, 1, 2], is used, so AS = 3 (AS denotes
Alphabet Size)

Important to note that the number of levels L is unlimited and it is not related to AS. Due to space constraints only 3 levels are shown (L =3)

The oval nodes contain the symbol associated with that node, while the square boxes next to the nodes depict the strings which are the concatenations of all symbols which were visited by arriving at that node.

Figure 1 :
Base-3 Complete Tree Structure Diagram

Stage 4b : Properties of Complete Tree Structures

1. They require zero storage (no memory management, rebalancing trees etc.)

2. Complex recursive navigation procedures are not required.

3. The design is extremely simple and elegant.

4a. Is very powerful, sports near-zero access time

4b. Presents unlimited virtual extent to users

5. Versatile, accommodates most of the fundamental data types such as integers, floats, strings, any integer base from binary to hexadecimal and beyond.

6 Possess a very important, fundamental 4-way equivalence :

==

6.1. The decimal integer counting numbers which use the no-leading-0, [0-9] alphabet

6.2. The inherent contents of each node, which is always one of the members of the defining alphabet of the complete tree ([0,1,2] in the example below .

6.3. The terminal contents of each node, which is the concatenation of all the node contents from the first level to the current node . For example for node 27, this is [1, 1, 2]
Each terminal content is enclosed in a rectangle. Note that for the first level, the inherent and terminal contents coincide. For node #2 it is 1 for both.

6.4. The set of terminal contents of each node is equivalent to the complete set of variations (combinations followed by permutations) of a set of its defining alphabets (such as symbols [0-9] for
decimal and [0-1] for binary or [0-9, "."] for decimal floats.
Note that the length of the variations ranges from 1 to Lmax, and the number of variations is N

6.5. It is very important to observe that the orderly generation of the entire set of variations demands strict discipline. One cardinal rule is that for a variation to be unique - that is counted only once - **it must be rooted in one of nodes residing on level 1.** These level-1 nodes will all have a single-character content (matching every symbol in the alphabet.)

Then from each level-1 node, we must generate all possible second level variations, which will contain 2 nodes. This is repeated once more for level 3, where each variation will be made up of a sequence of 3 nodes, from level 1, level 2 and level 3 in that order.

To generate novel variations (without duplicates) we have one of the 3 possible choices :
1 node : from level 1
2 nodes :one from level 1 and one from level 2, in that order
3 nodes : one from level 1, one from level 2 and one from level 3, in that order.

No other mix or sequencing of the nodes is allowed !
(below I list the invalid node combinations)

More than one node from any of the 3 levels,
exactly 1 node from level 2 or exactly 1 node from level 3,
2 nodes, one from level 1 and another from level 3,
2 nodes, one from level 2 and another from level 3.

These rules might seem a bit complicated (and draconian) at the beginning, but bear with me. In no time our attention to detail will pay fantastic dividends as they will enable us to manage a limitless virtual storage of any vocabulary which can be generated by arbitrary alphabets.

Shown in Figure 1 above, is one of the simplest examples, the base-3 alphabet consisting of 3 symbols : [0, 1, 2] Descending to binary would have flattened the complete tree structure to the point where demonstrating the features of the complete tree storage and retrieval would have been awkward at best.

This diagram of the ternary tree also serves to demonstrate the interdependence of counters addresses, single contents and contents formed by linking an arbitrary number of nodes within a complete tree.

1. The decimal integer counters are determining the breadth-first position of the nodes. That is we advance starting at the leftmost node of the top level (L=1) sweep to the right then descend to the next

level. This way each and every node in the complete tree will be tagged and counted.

2. The nodes are depicted using ovals. Each node contains one single character of the pre-defined alphabet : In the current case : alphabet = [0, 1, 2], so a node can contain either 0, 1 or 2.

3. The strings formed by concatenating the contents of nodes are constituting our story.
The construction of the story always taking place depth first, that is from one level to the next, using the pre-determined links which must be defined prior of utilizing the complete tree.

4. Stories are not explicitly stored. They formed by extracting the contents of the nodes in the order the nodes are accessed using the pre-determined links which fix the location of each node.

5. There are 3 important metrics of a tree :

5a. The depth of the tree which is the same as the number of levels constituting the tree.
   It is denoted by L. Each node must be residing in a level, no exceptions.
   (note that L may stand both for Level or for Length)

5b. The alphabet, which is a set of unique symbols with its length denoted by Alphabet Size or AS

See Comprehensive "Complete Tree" formulas at end of this chapter !!!

This implied association is absolutely serendipitous and whoever invented it was a genius.

At any rate, complete trees hand us a vast, unlimited, well-organized and easily accessible storage arrangement second to none for performing a very specific task of counting by comparing the contents of two equal sized instances of the corresponding complete trees.

For purposes of counting or streamlined generation of specific patterns this data structure is the most elegant one can ever invent. Throw in the incredible bonus that it takes zero actual storage space and you know that you are looking at one of the timeless gems of algorithmic par excellence. I can't stop praising it, as the complete tree structure made it possible for me to accomplish a task which at least for the past 6 generations was judged to be impossible : **the counting of floating point numbers.**

Stage 4c : Near-Complete Tree Structure or NCTS of decimal integer counters,
using the complete tree design features we introduced for the ternary and binary case above :

Figure 2 below demonstrates the essence of NCTS counting by beads. We see a particular structure which consists of identical size boxes labeled by a predetermined sequence of base-10 numerical symbols.

Within each box there is a single bead. (depicted by the rectangles) The first bead in the entire structure is in the box with the label of 1, the 9-th bead is box 9

Figure 2
Counting with Decimal Beads

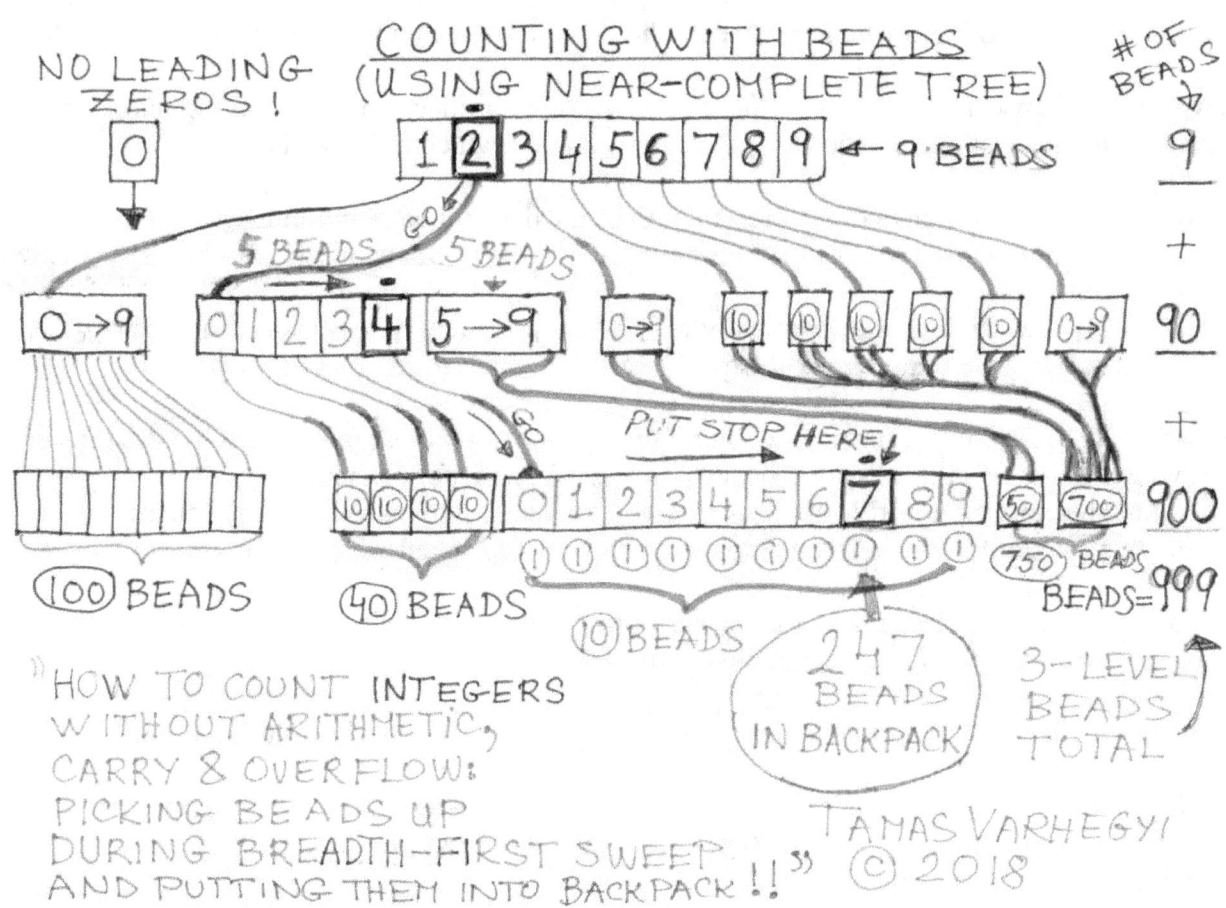

The example shows how to get the number of beads which correspond to the decimal number 247.

Step 1 is to locate the third digit of 247 on level 3 using depth-first navigation Once found, we put a marker a kind of stop sign there.

Next we traverse the tree in breadth-first sequence (horizontally, exhausting all nodes from level 1 to level 3, while moving from left to right on each level and collecting 1 bead from each node.

In the current example levels 1 and 2 are visited completely, but level 3 is traversed only up to the "stop-sign" which was placed at digit "7" as shown. If done right the pile of beads in our basket will represent the decimal number 247.

Now this is the time to be impressed. After all these details about tree traversal we actually have to do nothing.

The tree shown above does not represent actual memory locations, it only exist as a virtual navigation and counting aid to the person who designed the access algorithm traversing this tree.
(in this case myself)

Reinventing Cascaded Carry

Turns out that the digit sequences generated by traversing the near-complete tree in breadth first order will be nothing more and nothing less than the progression of the positive integer counting numbers, e.g., 1, 2, 3,...9, 10,...19,20,.....99,100, ...999,1000 and so on. The resulting sequences will be identical to the ones produced when we execute the **cascaded-carry** for integers. See table at Stage 3 !

We have shown the most important relationship between the primary integer counting numbers and the discrete physical items which are the underpinning of all counting activities.

From now on the near-complete tree representation of the decimal counting integers will be the anchor for counting tasks we plan to undertake on integers or floats of arbitrary bases.

However, please keep in mind that eventually we will confine ourselves to using only decimal counting agents which will make it unnecessary to employ complete or near-complete tree structures.

Most mathematical, scientific and engineering literature and computer applications have extensive utilities which automated the **cascaded-carry process.** As a matter of fact, it is one counting method which became ingrained in all of us from an early age. We mastered it to such a degree that we can do it in our head on autopilot without using calculators or computers.

As we have explored the mechanics of integer counting we eventually realized that the floating point numbers or floats for short are completely missing.

Integers are excellent for counting but they have a giant drawback : fixed resolution of unity.

So we will proceed the counting of floats as well, handing us mastery over another important domain which will allow access to a number system with limitless resolution by the introduction of negative powers of base.

However we are very lucky. The nearly unmanageable complexity of floats which are the unfortunate side-effect of having to deal with the negative powers may be sidestepped when counting them.

We do that by an ingenious trick. We implement the counting algorithms by using only the lexicographical and fake integer properties of the floats. That is, we will arrange the floats by interpreting both their character representation and incidental integer values.

I will clarify this later.

Footnote :

The differing real estate occupied by integer counters expressed in different number bases has no bearing on the one-to-one correspondence of such integers.

No one would dispute that the far more verbose binary representation and the decimal variety of the same quantities do not correspond to each other.

Chapter 5b

Complete Tree Structures Implementation

Stage 5: Proving the countability of floats using Complete Tree Structures

We have arrived at the one of the critical proofs which will neutralize all past and future claims about the uncountability of floats.

We will use two complete trees which share a couple of fundamental facts :

The 11-symbol [0-9, A] complete tree correctly produces all the integer **variations** of the 11-symbols, courtesy of the structure of the complete tree and the mechanics of traversing this tree. (we assume that the symbol A represents the integer 10)

The 11-symbol [0-9, dp] complete tree correctly produces all the **variations** of the 11-symbols containing the decimal point, therefore will produce all decimal candidate floats (and a very large number of gibberish).

The idea is truly simple. We will construct two nearly identical base-11 complete trees. As we have shown before, a complete tree is sufficiently defined by the number and the composition of its core alphabet.

1. Let us define two base-11 complete trees, with the alphabets specified as:
base_11_int_seqno = [0, 1, 2, 3, 4, 5, 6, 7, 8, 9, A] and
base_10_floats = [0, 1, 2, 3, 4, 5, 6, 7, 8, 9, dp]

(where dp stands for the period character [.] which serves as a decimal point for all base-10 floats.)

The two alphabet sizes are equal,   or AS(base_11_int_seq) = AS(base_10_floats) = 11

Now, without a formal proof I claim the following facts :

2. If two complete trees have the same number of symbols in their respective alphabets then regardless of the particular details of the symbols, they will have identical structures and all deposit/navigation and retrieval operations will be equivalent.

By traversing the trees in tandem we could produce pairs of node contents which will have a one-to-one relationship between them. In essence we will match each float candidate with an integer counting number.

3. The only difference will be that the strings retrieved will have dissimilar compositions of characters. However, the number and size distribution of the entire set of generated strings will be identical.

4. From this we conclude that every string which could be generated in one tree will have its counterpart in the other, level by level up to an arbitrarily large number of levels. Thus perfect one-to-one correspondence will be achieved.

5. The first complete tree will generate the correct base-11 integer counting agents (with leading zeros).

6. The second tree, "base_10_floats" contains the entire collection of all base-10 strings which can be

formed from the symbol set [0, 1, 2, 3, 4, 5, 6, 7, 8, 9, dp] . Let's call these strings **candidate floats**.

7. It has been claimed for centuries that …
It is not possible to count floats, and that the cardinality of floats far exceeds the cardinality of integers.

8. Contrary for those claims we have shown that :

8.a By using a base-11 integer counting agent it is possible and trivial to count candidate floats

8.b That for each candidate float which can be constructed there is a integer counting agent **n**.

8.c To use the official phrase : There is perfect **one-to-one correspondence** between the base-11 integer counters and the base-10 candidate floats.

8.d From the set of candidate floats we will be able to extract a subset which will correspond all possible legitimate floats.

9. Since not all base-10 candidate floats will qualify to be valid floats we will end up with leftover integer counters. This means that there will always be more integer counters than valid floats.

Conclusion : The candidate floats and even more the valid floats are eminently countable.

Figure 3
Using complete" base-11 counting agent

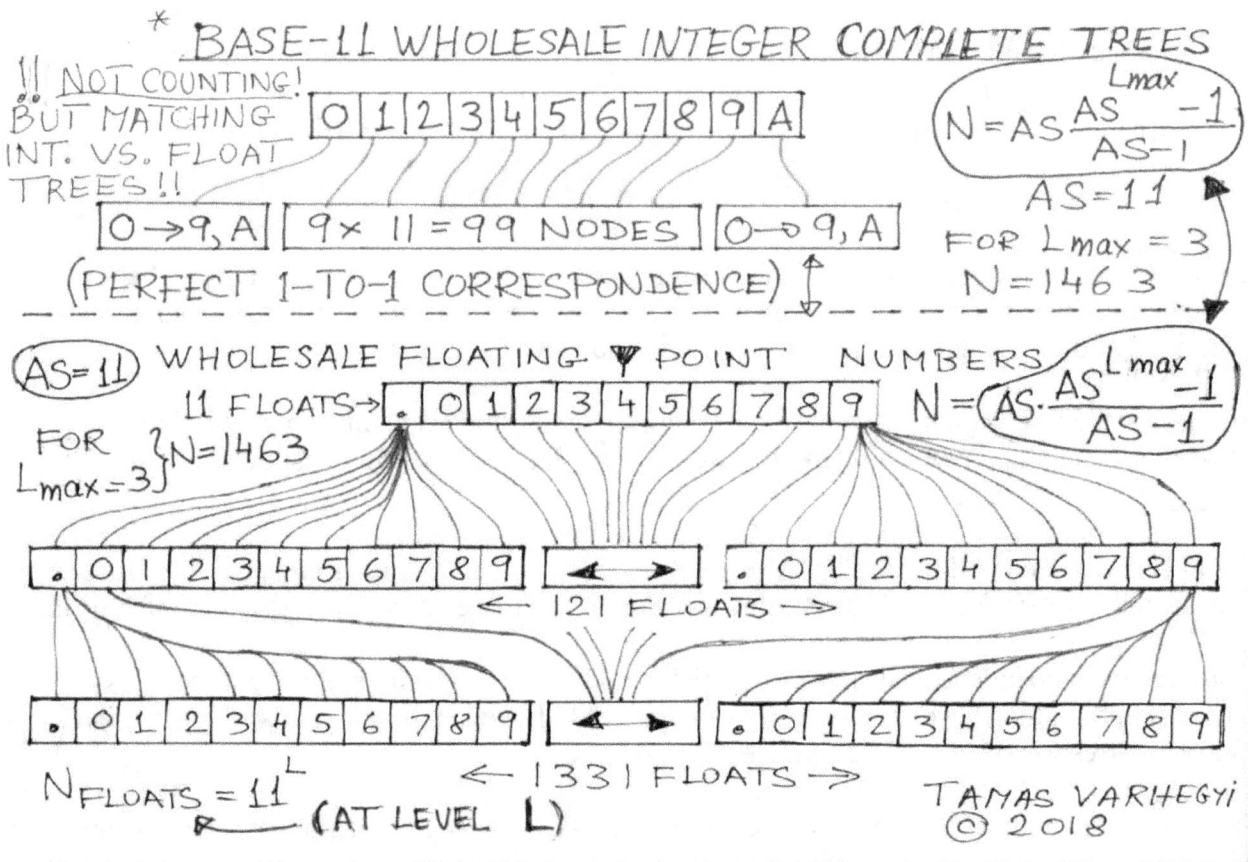

It is important to note that in the complete tree diagrams in Figure 3 we allow both a leading 0 on level 1 for the integer tree and a leading decimal point for the float tree - also on level 1. Both nodes could be removed without influencing the basic thrust of our argument.

However, this would force us to make a confession that we are no longer working with perfect trees and explain the reason and the impact (as we did here). The elegance of the proof which uses these two perfect complete trees would be somewhat diminished.

Leaving them in increases the total count by the factor of AS/(AS-1) which is inconsequential.

Note : Since we never count integers with leading zeros and certainly don't allow leading decimal points for floats so those nodes will have to be removed eventually.

We have tabulated partial results in Figure ?? Obviously there are a vast amount of invalid floats on the right, however by applying a float syntax filter they can be removed, to show a valid progression of floats we can generate.

Please note that the presence of invalid floats in the floats complete tree is one more argument for one-to-one correspondence, since there will be a good heap of failed candidate floats which will no longer use up precious counting agents.

Now we will be able to ratchet down the base of the integer counting agent tree a notch and make the following adjustments:

Stage 6 : Counting floats with a near-complete base-10 counting agent

1. We will switch to a base_10_int_seqno = [0, 1, 2, 3, 4, 5, 6, 7, 8, 9] alphabet

2. Even better we will use a "near-complete" base-10 counting agent. We will remove the [0] symbol from the first level of the complete tree. All previously present leading zeros will be removed from the generated counting agents. In essence we have returned to the quintessential counting numbers : the positive decimal integers, thank God, it was about time.

Please note that the leading 0 node on level 1 for the integer complete tree is removed. We must refer to it as a **near**-complete tree. But we left the leading decimal point node for the float tree on level 1.

Here is why. After we generate the level-by-level candidate floats (most of which will be gibberish) we can optionally apply a syntax check to extract valid floats only. This will be done below with the results tabulated.

Looking forward, no matter how clever we are in generating, then scrubbing candidate floats to arrive at a list of correct floats, the base-10 float tree (complete or near-complete makes no difference) will be a giant black hole sucking the last ounce of computer processing resources and our precious free time dry.

We will have to be near geniuses and find an alternative method, it will be hard but worth the effort.

See Figure 4 below :

Figure 4 Using a near-complete" base-10 counting agent

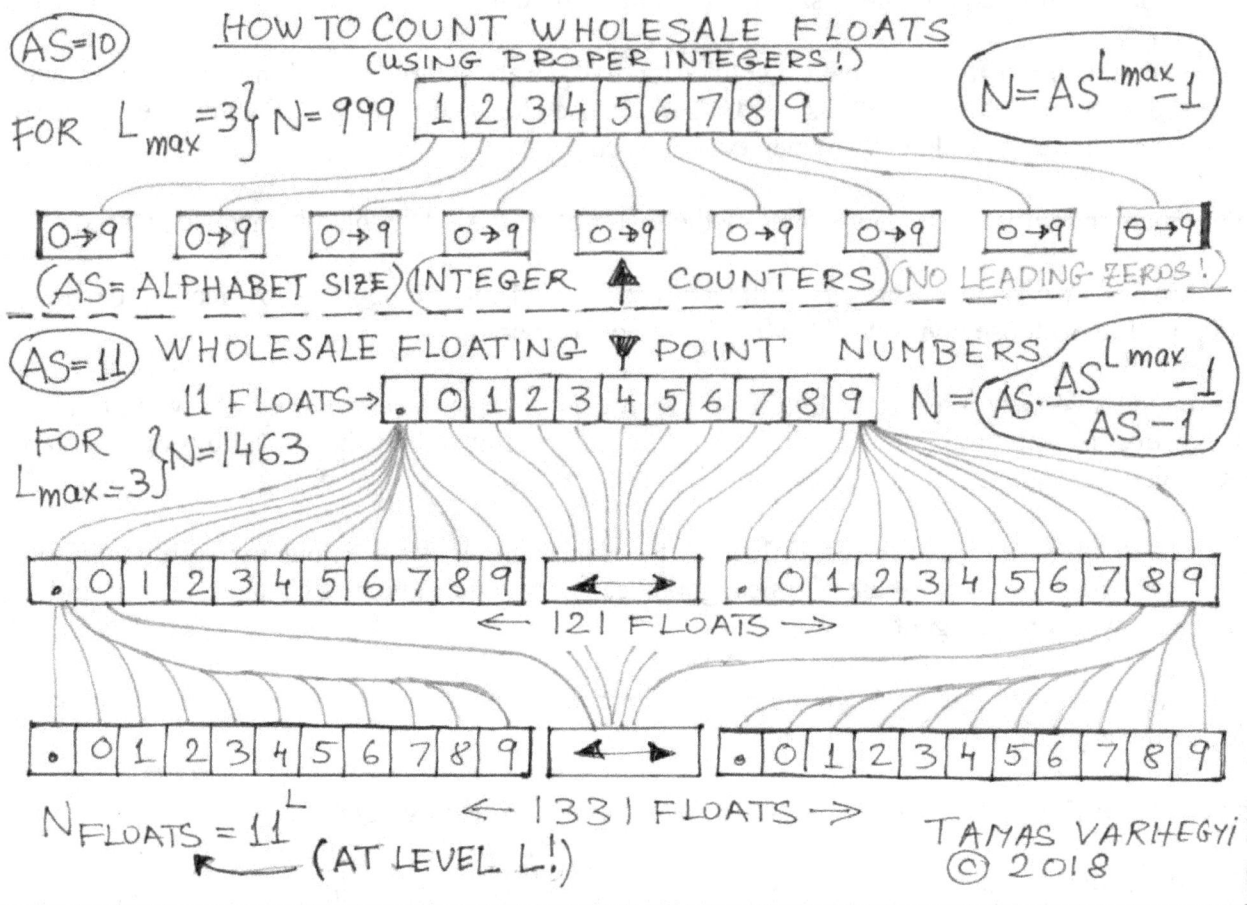

Before proceeding let us explore a truly powerful insight which will annihilate any claim
of a meaningful difference between the cardinalities of integers and floats.

We are already very familiar with the 2 alphabets, both accommodating decimal integers :
The first :   base_10_int_seqno = [0, 1, 2, 3, 4, 5, 6, 7, 8, 9]
The second : base_10_floats = whole_integers . fractional_integers = [0, 1, 2, 3, 4, 5, 6, 7, 8, 9, dp]

Let us replace the base_10_floats alphabet with another one, which will have two separators : $ and &

integers : base_10_int_seqno = [0, 1, 2, 3, 4, 5, 6, 7, 8, 9]
two_separators : integers_1 $ integers_2 & integers_3 =   [0, 1, 2, 3, 4, 5, 6, 7, 8, 9, $, &]

Example numbers of the two_separators alphabet : 4$7&55   or 11111$22222&33333
What we have just created is a 3-integer compartment arrangement being counted by
single-compartment integers.

If we now followed the Cantorian cardinalities we would have 3 of them : aleph1, aleph2 and aleph3,
with the attendant consequences of associating 3 infinite integer sets. So the two_separators
alphabet would be even more uncountable than the floats were (quoting Cantor and everybody else)

The argument would proceed as follows : For each number in integers_1 there would be an infinite
number in integers_2 and for every number in integers_2 there would be an infinite number in
integers_3.

A questionable analogy would be : N_TOTAL $\approx \infty * \infty * \infty$ a hopelessly uncountable monstrosity.

Do not despair. Let's remember the two_separators alphabet : [0, 1, 2, 3, 4, 5, 6, 7, 8, 9, $, &]

This alphabet is easily representable in a base-12 Complete Tree Structure, showcasing the symbols in the alphabet.

We then would have the choice of counting them using either a base-12 CTS or a base-10 decimal CTS. Either way, a 3-integer compartment of number representation (a kind of float on steroids still would be eminently countable).

This subject refuses to go away. It just dawned on me, why bother with two distinct separators? Just change the syntax rules and allow a variant of floats to have up to two decimal points. Then the float CTS would remain identical while allowing the 3-integer compartment variety of floats.

Now we are done for a while !

So what is the lesson learned? The most important one is that we must not quit if some counting method, most likely a boneheaded one returns a "no-can-do" verdict. We must always try to count the numbers by populating a CTS with the prevailing alphabet. Of course this will not be always possible, but for simple, linear alphabets with unique symbols this is the way to go.

Summary of previous actions :

We have arrived here after we mastered the generation of a complete set of variations of a base-11 alphabet.

Specifically we have used the set : [0,1,2,3,4,5,6,7,8,9,dp] to obtain 1331 candidates from which to extract those 3-character variations which obey the strict rules we devised for a proper float. [see Chapter 3b : Structure and Syntax Rules of Integers and Floats

We are almost there. We have the ability to generate raw variations of which contain all eligible floats of a certain character length and the syntax rules which will select only strings which pass the syntax check.

The next step was to integrate the various algorithms into a set of Maple procedures which now can :

1. Generate the raw base-11 variations from a user supplied arbitrary alphabet.

2. Select the variations which pass the syntax check

3. Output the set of valid floats of a given character length.

4. Allow some user input as to the character length and the range of floats desired.

Figure 5
Raw float output of decimal floats [. 0 -> 9] -using

TOP OF PAGE !

BOTH VALID AND GIBBERISH FLOAT STORIES WERE PRODUCED (BUT NO VALID FLOATS ARE DISPLAYED HERE

BASE-11, 3-LEVEL COMPLETE TREE.

COMPLETE THREE-CHARACTER VARIATIONS OF THE ALPHABET: [.,0,1,2,3,4,5,6,7,8,9]

TAMAS VARHEGYI © 2018

A SUBSEQUENT FILTERING ALGORITHM WILL PRODUCE EXACTLY 90 VALID 3-CHARACTER FLOATS!!

11	9
10	8
9	7
8	6
7	5
6	4
5	3
4	2
3	1
2	0
1	•

132	99		1463	999
131	98		1462	998
130	97		1461	997
129	96		1460	996
128	95		1459	995
127	94		1458	994
126	93		1457	993
125	92		1456	992
124	91		1455	991
123	90		1454	990
122	9•		1453	99•
121	89		1452	989
56	3•		562	25•
55	29		561	249
54	28		560	248
53	27		559	247
52	26		558	246
51	25		557	245
50	24		556	244
49	23		555	243
48	22		554	242
47	21		553	241
46	20		552	240
45	2•		551	24•
44	19		550	239
23	0•		144	•0•
22	•9		143	••9
21	•8		142	••8
20	•7		141	••7
19	•6		140	••6
18	•5		139	••5
17	•4		138	••4
16	•3		137	••3
15	•2		136	••2
14	•1		135	••1
13	•0		134	••0
12	••		133	•••

11 TOTAL   121 TOTAL   1331 TOTAL

GRAND TOTAL = 11+121+1331 = 1463 FLOATS

Figure 6
Raw float output of decimal floats [. 0 -> 9] -using
near complete (no leading zeros) base-11 integer counters [0 -> A]

(only the last 3 floats are valid !!! see bolded, lower right corner)

1 = "."	40 = "19"	7A = 58"	109 = "97"	148 = ".26"	187 = ".65"
2 = "0"	41 = "2."	80 = "59"	10A = "98"	149 = ".27"	188 = ".66"
3 = "1"	42 = "20"	81 = "6."	110 = "99"	14A = ".28"	189 = ".67"
4 = "2"	43 = "21"	82 = "60"	111 = "..."	150 = ".29"	18A = ".68"
5 = "3"	44 = "22"	83 = "61"	112 = "..0"	151 = ".3."	190 = ".69"
6 = "4"	45 = "23"	84 = "62"	113 = "..1"	152 = ".30"	191 = ".7."
7 = "5"	46 = "24"	85 = "63"	114 = "..2"	153 = ".31"	192 = ".70"
8 = "6"	47 = "25"	86 = "64"	115 = "..3"	154 = ".32"	193 = ".71"
9 = "7"	48 = "26"	87 = "65"	116 = "..4"	155 = ".33"	194 = ".72"
A = "8"	49 = "27"	88 = "66"	117 = "..5"	156 = ".34"	195 = ".73"
10 = "9"	4A = "28"	89 = "67"	118 = "..6"	157 = ".35"	196 = ".74"
11 = ".."	50 = "29"	8A = 68"	119 = "..7"	158 = ".36"	197 = ".75"
12 = ".0"	51 = "3."	90 = "69"	11A = "..8"	159 = ".37"	198 = ".76"
13 = ".1"	52 = "30"	91 = "7."	120 = "..9"	15A = ".38"	199 = ".77"
14 = ".2"	53 = "31"	92 = "70"	121 = ".0."	160 = ".39"	19A = ".78"
15 = ".3"	54 = "32"	93 = "71"	122 = ".00"	161 = ".4."	1A0 = ".79"
16 = ".4"	55 = "33"	94 = "72"	123 = ".01"	162 = ".40"	1A1 = ".8."
17 = ".5"	56 = "34"	95 = "73"	124 = ".02"	163 = ".41"	1A2 = ".80"
18 = ".6"	57 = "35"	96 = "74"	125 = ".03"	164 = ".42"	1A3 = ".81"
19 = ".7"	58 = "36"	97 = "75"	126 = ".04"	165 = ".43"	1A4 = ".82"
1A = ".8"	59 = "37"	98 = "76"	127 = ".05"	166 = ".44"	1A5 = ".83"
20 = ".9"	5A = "38"	99 = "77"	128 = ".06"	167 = ".45"	1A6 = ".84"
21 = "0."	60 = "39"	9A = "78"	129 = ".07"	168 = ".46"	1A7 = ".85"
22 = "00"	61 = "4."	A0 = "79"	12A = ".08"	169 = ".47"	1A8 = ".86"
23 = "01"	62 = "40"	A1 = "8."	130 = ".09"	16A = ".48"	1A9 = ".87"
24 = "02"	63 = "41"	A2 = "80"	131 = ".1."	170 = ".49"	1AA = ".88"
25 = "03"	64 = "42"	A3 = "81"	132 = ".10"	171 = ".5."	200 = ".89"
26 = "04"	65 = "43"	A4 = "82"	133 = ".11"	172 = ".50"	201 = ".9."
27 = "05"	66 = "44"	A5 = "83"	134 = ".12"	173 = ".51"	202 = ".90"
28 = "06"	67 = "45"	A6 = "84"	135 = ".13"	174 = ".52"	203 = ".91"
29 = "07"	68 = "46"	A7 = "85"	136 = ".14"	175 = ".53"	204 = ".92"
2A = "08"	69 = "47"	A8 = "86"	137 = ".15"	176 = ".54"	205 = ".93"
30 = "09"	6A = "48"	A9 = "87"	138 = ".16"	177 = ".55"	206 = ".94"
31 = "1."	70 = "49"	AA = "88"	139 = ".17"	178 = ".56"	207 = ".95"
32 = "10"	71 = "5."	100 = "89"	13A = ".18"	179 = ".57"	208 = ".96"
33 = "11"	72 = "50"	101 = "9."	140 = ".19"	17A = ".58"	209 = ".97"
34 = "12"	73 = "51"	102 = "90"	141 = ".2."	180 = ".59"	20A = ".98"
35 = "13"	74 = "52"	103 = "91"	142 = ".20"	181 = ".6."	210 = ".99"
36 = "14"	75 = "53"	104 = "92"	143 = ".21"	182 = ".60"	211 = "0.."
37 = "15"	76 = "54"	105 = "93"	144 = ".22"	183 = ".61"	212 = "0.0"
38 = "16"	77 = "55"	106 = "94"	145 = ".23"	184 = ".62"	**213 = "0.1"**
39 = "17"	78 = "56"	107 = "95"	146 = ".24"	185 = ".63"	**214 = "0.2"**
3A = "18"	79 = "57"	108 = "96"	147 = ".25"	186 = ".64"	**215 = "0.3"**

The results for all 3character floats and a subset of the 4-character floats are shown in the table below.

Figure 7
List of floats for Level #1 and partial Level #2 up to 1.99 = 270

©

1 0.1	2 0.2	3 0.3	4 0.4	5 0.5	6 0.6	7 0.7	8 0.8	9 0.9
10 1.1	11 1.2	12 1.3	13 1.4	14 1.5	15 1.6	16 1.7	17 1.8	18 1.9
19 2.1	20 2.2	21 2.3	22 2.4	23 2.5	24 2.6	25 2.7	26 2.8	27 2.9
28 3.1	29 3.2	30 3.3	31 3.4	32 3.5	33 3.6	34 3.7	35 3.8	36 3.9
37 4.1	38 4.2	39 4.3	40 4.4	41 4.5	42 4.6	43 4.7	44 4.8	45 4.9
46 5.1	47 5.2	48 5.3	49 5.4	50 5.5	51 5.6	52 5.7	53 5.8	54 5.9
55 6.1	56 6.2	57 6.3	58 6.4	59 6.5	60 6.6	61 6.7	62 6.8	63 6.9
64 7.1	65 7.2	66 7.3	67 7.4	68 7.5	69 7.6	70 7.7	71 7.8	72 7.9
73 8.1	74 8.2	75 8.3	76 8.4	77 8.5	78 8.6	79 8.7	80 8.8	81 8.9
82 9.1	83 9.2	84 9.3	85 9.4	86 9.5	87 9.6	88 9.7	89 9.8	90 9.9

Figure 7 continued

3-character floats above (90 out of 90 shown)
4 character floats below (180 out of 1710 shown)

91 0.01	92 0.02	93 0.03	94 0.04	95 0.05	96 0.06	97 0.07	98 0.08	99 0.09
100 0.11	101 0.12	102 0.13	103 0.14	104 0.15	105 0.16	106 0.17	107 0.18	108 0.19
109 0.21	110 0.22	111 0.23	112 0.24	113 0.25	114 0.26	115 0.27	116 0.28	117 0.29
118 0.31	119 0.32	120 0.33	121 0.34	122 0.35	123 0.36	124 0.37	125 0.38	126 0.39
127 0.41	128 0.42	129 0.43	130 0.44	131 0.45	132 0.46	133 0.47	134 0.48	135 0.49
136 0.51	137 0.52	138 0.53	139 0.54	140 0.55	141 0.56	142 0.57	143 0.58	144 0.59
145 0.61	146 0.62	147 0.63	148 0.64	149 0.65	150 0.66	151 0.67	152 0.68	153 0.69
154 0.71	155 0.72	156 0.73	157 0.74	158 0.75	159 0.76	160 0.77	161 0.78	162 0.79
163 0.81	164 0.82	165 0.83	166 0.84	167 0.85	168 0.86	169 0.87	170 0.88	171 0.89
172 0.91	173 0.92	174 0.93	175 0.94	176 0.95	177 0.96	178 0.97	179 0.98	180 0.99
181 1.01	182 1.02	183 1.03	184 1.04	185 1.05	186 1.06	187 1.07	188 1.08	189 1.09
190 1.11	191 1.12	192 1.13	193 1.14	194 1.15	195 1.16	196 1.17	197 1.18	198 1.19
199 1.21	200 1.22	201 1.23	202 1.24	203 1.25	204 1.26	205 1.27	206 1.28	207 1.29
208	209	210	211	212	213	214	215	216

1.31	1.32	1.33	1.34	1.35	1.36	1.37	1.38	1.39
217	218	219	220	221	222	223	224	225
1.41	1.42	1.43	1.44	1.45	1.46	1.47	1.48	1.49
226	227	228	229	230	231	232	233	234
1.51	1.52	1.53	1.54	1.55	1.56	1.57	1.58	1.59
235	236	237	238	239	240	241	242	243
1.61	1.62	1.63	1.64	1.65	1.66	1.67	1.68	1.69
244	245	246	247	248	249	250	251	252
1.71	1.72	1.73	1.74	1.75	1.76	1.77	1.78	1.79
253	254	255	256	257	258	259	260	261
1.81	1.82	1.83	1.84	1.85	1.86	1.87	1.88	1.89

3-character floats above (90 out of 90 shown)
4 character floats below (180 out of 1710 shown)

91 0.01	92 0.02	93 0.03	94 0.04	95 0.05	96 0.06	97 0.07	98 0.08	99 0.09
100 0.11	101 0.12	102 0.13	103 0.14	104 0.15	105 0.16	106 0.17	107 0.18	108 0.19
109 0.21	110 0.22	111 0.23	112 0.24	113 0.25	114 0.26	115 0.27	116 0.28	117 0.29
118 0.31	119 0.32	120 0.33	121 0.34	122 0.35	123 0.36	124 0.37	125 0.38	126 0.39
127 0.41	128 0.42	129 0.43	130 0.44	131 0.45	132 0.46	133 0.47	134 0.48	135 0.49
136 0.51	137 0.52	138 0.53	139 0.54	140 0.55	141 0.56	142 0.57	143 0.58	144 0.59
145 0.61	146 0.62	147 0.63	148 0.64	149 0.65	150 0.66	151 0.67	152 0.68	153 0.69
154 0.71	155 0.72	156 0.73	157 0.74	158 0.75	159 0.76	160 0.77	161 0.78	162 0.79
163 0.81	164 0.82	165 0.83	166 0.84	167 0.85	168 0.86	169 0.87	170 0.88	171 0.89
172 0.91	173 0.92	174 0.93	175 0.94	176 0.95	177 0.96	178 0.97	179 0.98	180 0.99
181 1.01	182 1.02	183 1.03	184 1.04	185 1.05	186 1.06	187 1.07	188 1.08	189 1.09
190 1.11	191 1.12	192 1.13	193 1.14	194 1.15	195 1.16	196 1.17	197 1.18	198 1.19
199 1.21	200 1.22	201 1.23	202 1.24	203 1.25	204 1.26	205 1.27	206 1.28	207 1.29
208 1.31	209 1.32	210 1.33	211 1.34	212 1.35	213 1.36	214 1.37	215 1.38	216 1.39
217 1.41	218 1.42	219 1.43	220 1.44	221 1.45	222 1.46	223 1.47	224 1.48	225 1.49
226 1.51	227 1.52	228 1.53	229 1.54	230 1.55	231 1.56	232 1.57	233 1.58	234 1.59
235 1.61	236 1.62	237 1.63	238 1.64	239 1.65	240 1.66	241 1.67	242 1.68	243 1.69
244 1.71	245 1.72	246 1.73	247 1.74	248 1.75	249 1.76	250 1.77	251 1.78	252 1.79
253 1.81	254 1.82	255 1.83	256 1.84	257 1.85	258 1.86	259 1.87	260 1.88	261 1.89
262 1.91	263 1.92	264 1.93	265 1.94	266 1.95	267 1.96	268 1.97	269 1.98	270 1.99

Chapter 6a

Constructing Level_&_Bracket pyramids from float Complete Trees

Stage 7 : The big story : Floats must be ordered representation size (RS) first, value last.

1. The right way to order integers

It is well-known and universally accepted that integers are counted from smallest value toward larger ones. What is also obvious that the values increase in tandem with the representation sizes (RS) of the integers. Well, this is a somewhat sloppy definition. What we mean is that a representation size increase always forces a value size increase. For example :
RS = 1 for 1, 2, 3, 4, 5, 6, 7, 8, 9, then RS = 2 between 10 and 99, and RS = 3 for 100-999 and so on.

Why we can't order floats the same way ? After all when we count integers themselves we start with the smallest value 0 and increment steadily by 1 as long as we want to. But for integers, by its ingenious design the smallest values are represented by the shortest strings. 0 – 9 have 1 digit, 10->99 2 digits and so on. In case of the floats no such luck.

Floats have a built-in negative powers-of-ten association with the digits to the right of the decimal point. This results in a bizarre (well at least unusual) inverse relationship between inherent values and character sizes. For example 0.1 represents 10 billion times larger value than its far longer character string counterpart of 0.0000000001. Even worse, this inverse relationship cannot be relied on all the time as 10000000.1 has a larger value than 1.1

Let's have a look at it from a different perspective. Positive integers have a minimum value but positive floats do not, so there is no way to start counting by value. We must start at the smallest character size which for valid syntax floats happens to be 3. Then we go from minimum values to maximum values which can be accommodated by a given representation size. This sequencing gives us the glorious first valid float of [0.1].

Yes, we will take a hit as the values represented by the floats will oscillate, going up and down. We can't have it both ways but having a sound plan will serve us exceptionally well. The oscillating float values during counting will ensure that we make steady and fair accounting of both increasingly small and large floats.

2.a The right way to order floats is by using increasing size of representation then by value

For floats the values may increase or decrease with the representation size.
For example :   0.1, 0.01, 0.001, ... , The values decline while RS increases or :
            1.1, 10.1, 100.1, ...,   The values and RS both increase
            0.001, 0.01, 0.1, ... , The values increase while RS declines or :
            100.1, 10.1, 1.1, ...,   The values and RS both decline

Here we have to make a strange decision !!!

2.b The negative powers-of-10 association of the floats will be ignored !

Floats will be treated as either strings or a so-called fake integers in the algorithms we implement. Note that **fake integers** are defined in Chapter 3.

Also for this project I call **compatible floats** which have the same string length and a matching decimal point position, for example 13.42 and 99.57 are compatible. and the number of floats between them is computed by $9957 - 1342 + 1 = 8616$, which include the two barrier floats.

Then, the only way to count floats is to proceed from the smallest representation sizes (RS) to larger ones, while within a particular RS, we must count from the smallest value floats toward larger ones. As we do that we will be forced to step the floats through all the possible decimal point positions from 1 to RS-2.

Example : The shortest floats have RS = 3, so the first float is 0.1 (has the minimum value) and the last one whose RS=3 has 9.9 for its value. Next, for RS = 4, we have floats ranging from the lowest value of 0.01 to the highest 9.99 . Then we must shift the decimal point to the right and continue counting from 10.1 to 99.9 and so on. (Note that 00.1 is not a valid float !)

2.c Fake integers to the rescue

Just what do we mean it this case when we speak about the counting by fake integers ? Let's count from 10.1 to 99.9 If we retained the negative powers of 10 association of the floats it would be valid for us to try to go : 10.1, 10.11, 10.111 and so on. **Obviously we would never get there !!!**

By switching to fake integers the counting proceeds from 101 to 999 using the regular positive integer values in increasing order or : 101, 102, ... 998, 999 There is no possibility to get sidetracked into a dead-end corner. This is phenomenal.
If we want to know how many floats are between **two compatible** floats 10.1 and 99.9 inclusive, we simply take the difference of the two fake integers and add 1 : 999-101 = 898, so there are 898+1 = 899 floats between 10.1 and 99.9 inclusive.
[Note that for floats which are not compatible in the sense as we defined them above, counting the number of floats between them is not practical. As we will see attempts at manual counting will become unnecessary as there are appropriate utilities which will do the counting.]

By utilizing fake integers we not only can count floats but do so unambiguously, with ease and without limits. We will discuss this method of counting in much greater detail.

This is a good example of both insight and luck. It is certainly a truly novel way of counting the floats, but before I congratulate myself too much let's consider this :

The method we just discussed and propose to implement is not just the best way, it is the ONLY WAY! Thus anybody who decided to pursue the dream of counting the floats would have had to come to the identical conclusion.

The wholesale base-10 float tree we deployed in Chapter 5b already has the "character-size first" access bias by virtue of its design . This becomes obvious when we traverse the tree breadth first, depth second (sweep horizontally on level 1, then descend to level 2) and construct the float by extracting and concatenating the implied contents of the visited nodes. The smallest size "candidate floats" are generated first, followed by sizes 2, 3,...,n as we descend through the tree.

Stage 8 : Transforming the base-10 complete tree for float candidates into a vertical array structure containing only valid floats

(actually the base-10 complete tree for floats is an 11-symbol complete tree)

We execute the following steps :

3.1. Simplify the complete tree by depicting only the nodes which contain valid floats

3.2 Transform the near-complete reduced level-by-level float tree structure to a
virtual place-value vertical array structure (or place-value structure or **PVS** for short)
an incredibly simple arrangement where only valid floats reside.

3.3 From the **PVS** derive the L_&_B structures (**LBS** for short) and relationships
housing both floats and integer sequence numbers.

3.4 Design an efficient set of formulas which compute the beginning, size and end of brackets.

3.5 Develop two procedures to go from float to int_seqno and vice-versa from int_seqno to floats

3.6 Demonstrate actual conversions in both directions.

What if we could create an offspring of the complete trees, where no traversing takes place at all and both the integers and floats have their shared unique locations? These locations must be accessible both for deposits, referencing and retrieval (although deposits really do not take place for purposes other than visual displays.)

Transformations 3.1 and 3.2 are demonstrated below see Figures 6.1 and 6.2 below

You will notice shortly that the artwork in this book is not exactly sharp and professional.
Guilty as charged – severe time constraints and personal circumstances forced me to prioritize,
sharp and enticing graphics went to the bottom.

However the next illustration was deliberately made third rate. At this point in our quest to generate data structures from which some clever algorithms could extricate the seemingly elusive list of valid floats seemed to be next to impossible.

But I saw the light at the end of the tunnel and in a moment of weakness I yielded to the temptation to make the discovery of the perfect data structure as dramatic as possible. Painting a discombobulated, hopeless tangle of nodes with some valid floats trying to come up for air screaming to be taken seriously provided a stark contrast. Let's see if my plan worked or not.

At any rate, from this hopeless tangle below, we take the first step toward order by retaining only those nodes and the paths connecting them which terminate in valid floats.

Figure 6.1

Base-10 complete tree for float candidates – A true challenge

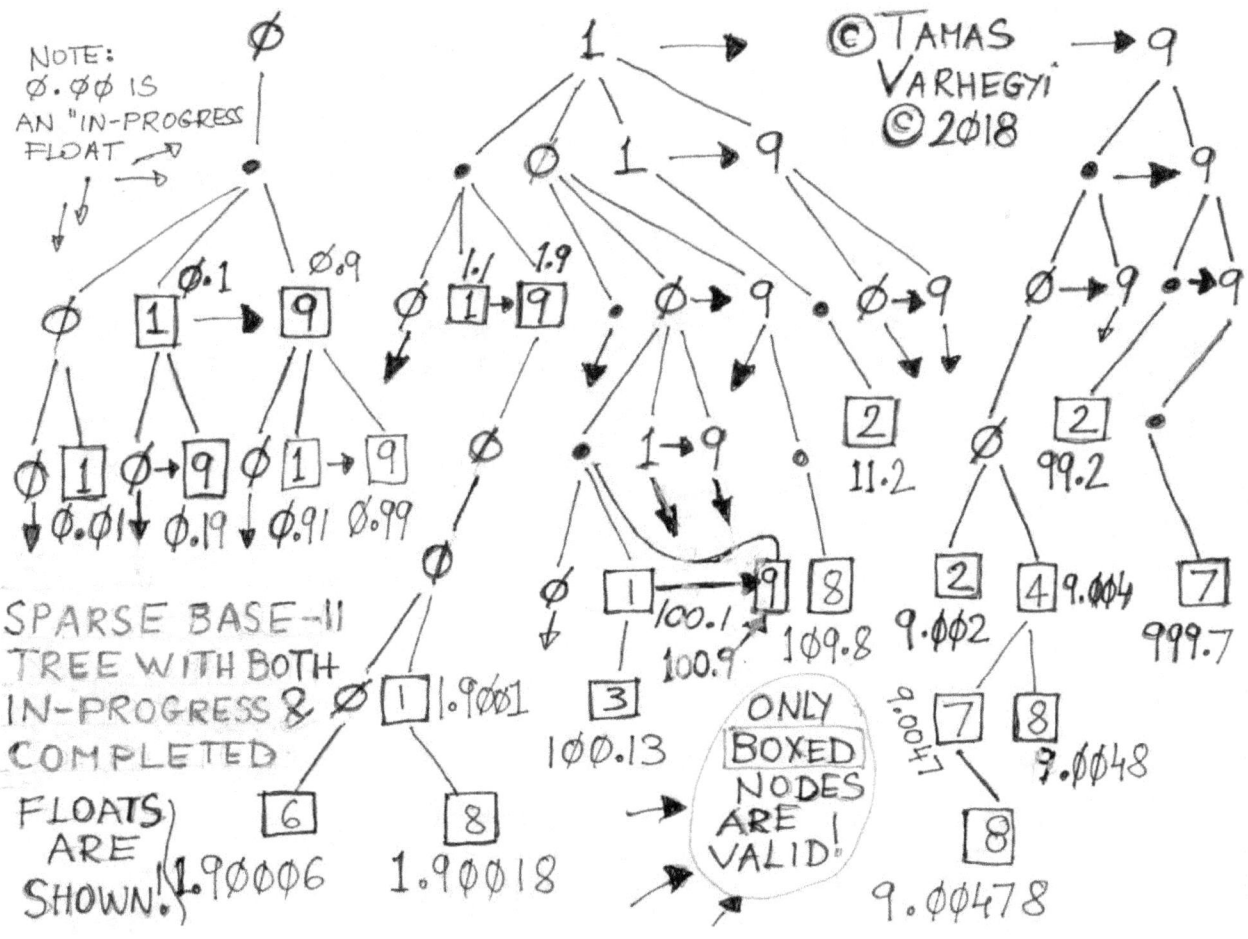

Before we continue, let's duplicate the 5-item declaration of the syntax rules as it will guide us in executing the conversion process :

1. Only the 10 digits [0,1,2,3,4,5,6,7,8,9] and the decimal point character ["."] may be used
2. The decimal point ["."] must be used exactly once.
3. No trailing or leading decimal points are permitted.
4. Trailing zeros are prohibited.
5. A leading zero is allowed only if it is immediately followed by a decimal point.

The left side of Figure 6.2 below shows the glorious transformation from a dead-end tangle to an orderly progression of the sub-trees we managed to create.

The vast, unmanageable single-root and all-connected tangle is behind us. We have created islands of order and simplicity. Each such entity is shared by floats which have identical character size (the Level or L) and decimal point position (which is referred to as the bracket_id or BR).

For example at L = 3, BR = 1, we collect all the floats from 0.1 to 9.9. They are all size 3 and each has a dp in the second position. For L = 4 we have two brackets : for BR = 1 the dp is in the second char position e.g. 1.34 and for BR = 2 the dp shifts to the right, then for example we have the float of 23.7.

Figure 6.2

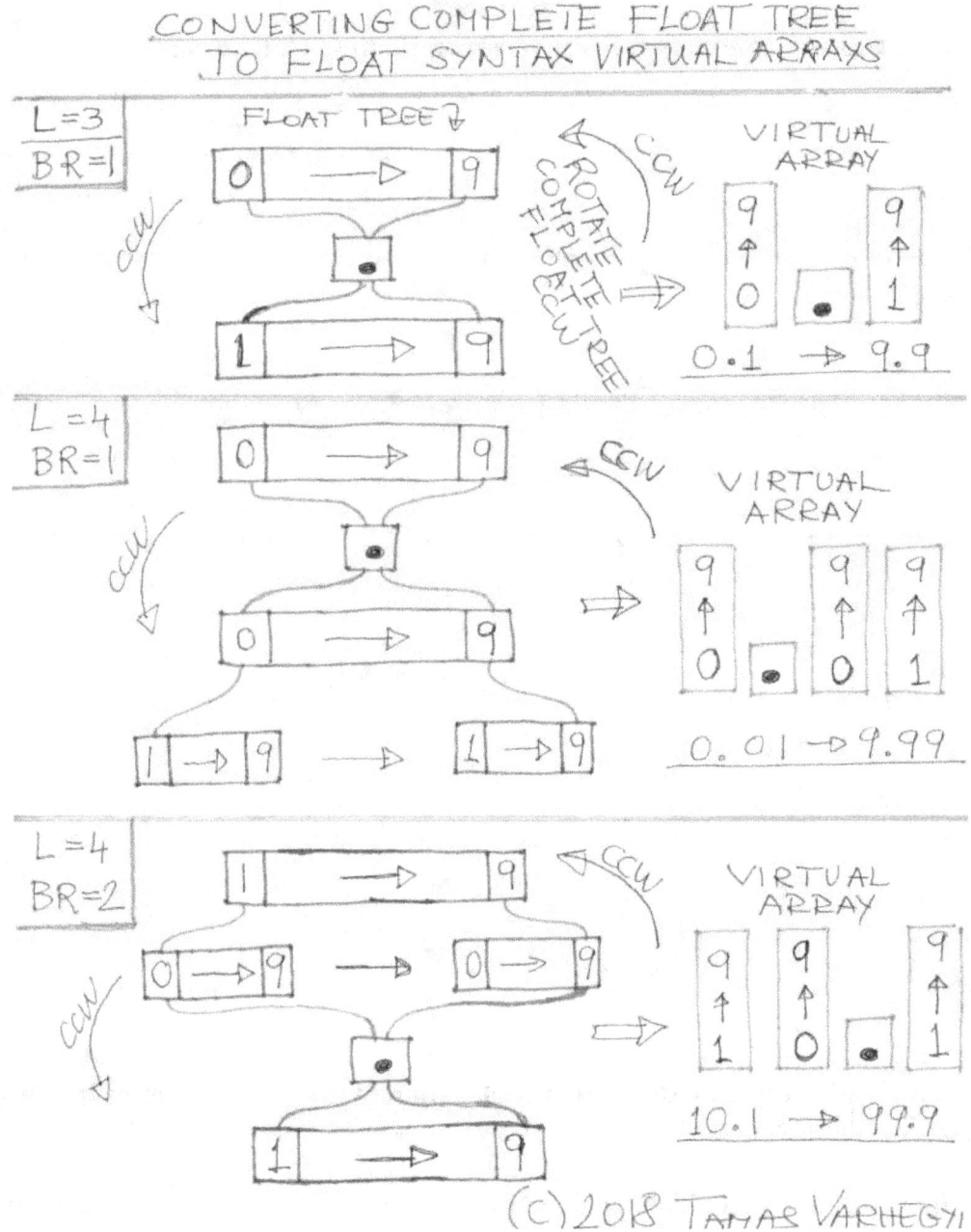

CONVERTING COMPLETE FLOAT TREE
TO FLOAT SYNTAX VIRTUAL ARRAYS

(C) 2018 TAMAS VARHEGYI

Incredibly enough these L_&_B mini-structures perform double duty : The exclude all invalid floats and accommodate all possible valid floats one can ever construct. Yet, we still have a software management nightmare.

These L&B structures as we mentioned before are "islands". As such they keep proliferating in an ever expanding pyramid-wave each having its internal structure with connected nodes, but the structures themselves are not connected to each other in any obvious way. If they were, we don't want to know as we are hell-bent on eliminating all links anyway whose management at this point appears next to impossible.

Luckily we have one more truly effective transformation to bail us out. We rotate each and every stand-alone float tree CCW or counter-clockwise and in addition remove all links. See Fig. 6.2 above

Vertical Place-Value Oriented (PVO) structures. (see Figure 6.3 next page)

The (PVO) structures which will readily yield to the orderly generation of valid floats - sequences of 10 digits and one separator obeying the 5 syntax rules.

This final virtual data structure matches the underlying logic of the character-oriented floats we use every day.

No more trees, no more links no more duplications. Just an exceptionally straightforward and dependable way of generating any valid float of unlimited value and character size.

The PVO's are organized in a pyramid where the grand total of the Levels & Brackets contain every possible valid float in an organized fashion.

That was our goal. See the right side of Figure 6.2. above

Table 1 : Level _&_ Bracket Pyramids for Base-10 PVO structures
(only float_stories are shown)

Lev=3 ["0.1" ->0 "9.9"]
br_id = 1

Lev=4 ["0.01" -> "9.99", "10.1" -> "99.9"]
br_id = 1   br_id = 2

Lev=5 ["0.001" -> "9.999", "10.01" -> "99.99", "100.1" -> "999.9"]
br_id = 1   br_id = 2   br_id = 3

Lev=6 ["0.0001" -> "9.9999", "10.001" -> "99.999", "100.01" -> "999.99", "1000.1" -> 9999.9"]
br_id = 1   br_id = 2   br_id = 3   br_id = 4

Figure 6.3

Vertical Place Value Oriented (PVO) structures.
(Derived from the right side of Figure 6.2)

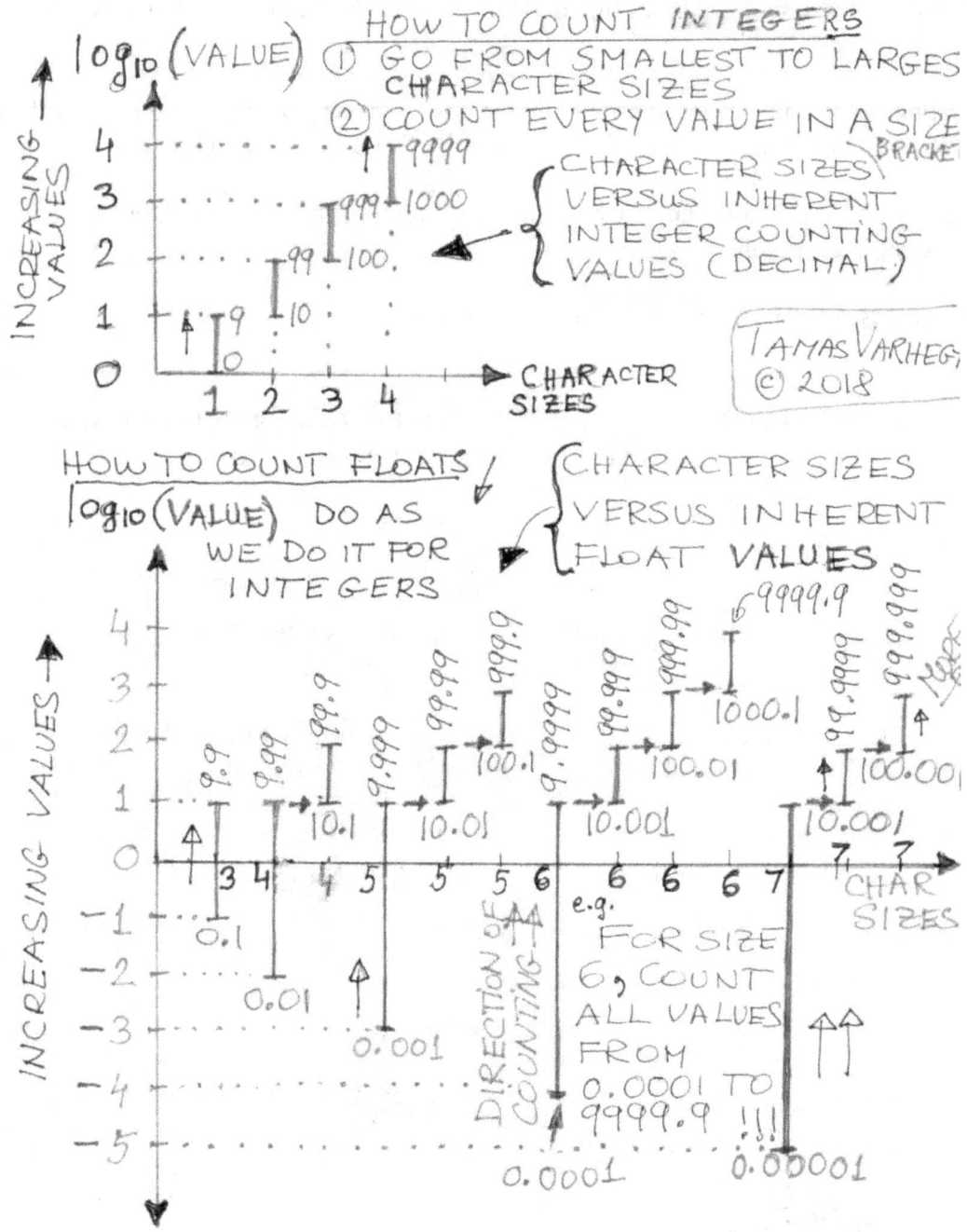

Stage 9 : Association between L&B pyramids and the decimal integer counters

We can observe that the L&B pyramid only contains easily generated floats. If we want to superimpose positive counting integers we need a pattern as we don't fancy brute force enumeration. Yes, eventually we could come up with a pattern but there is a far easier and more elegant way.

This is the moment of another serendipities discovery. Since we ultimately want to count character based floats (going as far as to call them strings) we can draw the following diagram. As you will see, the progression of the counting integers follows a recognizable pattern which we can obtain just by guessing. And we are done, this is certainly a nice bonus.

Simply by examining the structure we were able to associate a positive integer counting agent with each and every float in the structure. Thus each location has two occupants, a float and an integer sequence number. It is this dual representation which was our ultimate goal. Now let's see if are able to make the most of it. The goal is to write a pair of utilities which can do the following :

1a. Specify any float and obtain its location in the L&B pyramid.
1b. Use the L&B value to compute the value of the integer sequence number which also resides there.
2a. Specify any positive integer and obtain its location in the L&B pyramid.
2b. Use the L&B value to compute the value of the float which also resides there.

The details will be hidden in a twin set of Maple procedures, where a single call will produce either a float or an integer sequence number.

Table 2 : Integer L_&_B pyramid derived from the base-10 PVO structures
(only integer counters are shown)

L = 3 [1 -> 90] size = 90
br_id = 1

L = 4 [91 -> 990], [991 -> 1800] size =1710
br_id = 1   br_id = 2

L = 5 [1801 -> 10800], [10801 -> 18900], [18901 -> 27000] size =25200
br_id = 1   br_id = 2   br_id = 3

L = 6 [27001 -> 117000], [117001 -> 198000], [198001 -> 279000], [279001 -> 360000]
br_id = 1   br_id = 2   br_id = 3   br_id = 4

Table 3 : L_&_B pyramid : both int_seqno's and float_stories are shown

L=3 [I "=", 1, 90]
[F "=", "0.1", "9.9"]

L=4 [I "=", 91, 990, 991, 1800]
[F "=", "0.01", "9.99", "10.1", "99.9"]

L=5 [I "=", 1801, 10800, 10801, 18900, 18901, 27000]
[F "=", "0.001", "9.999", "10.01", "99.99", "100.1", "999.9"]

L=6 [I "=", 27001, 117000, 117001, 198000, 198001, 279000, 279001,360000]
[F "=", "0.0001", "9.9999", "10.001", "99.999", "100.01", "999.99", "1000.1", "9999.9"]

Consider the next two tables (3.a and 3.b) as our victory lap.

Their contents serve as a stunning demonstration between past disorder and disorientation.
In the first table 3.a, it is impossible to make sense of the progression of floats.

The problem here is that we order the floats according to inherent value sizes which make the representation sizes (character length) fluctuate wildly.

The seemingly random float – to - integer counting agents correspondence
Table 3.a

Case #	Float value ranges 0.0	Float strings in order of increasing values	Integer sequence numbers
1		"0.000000001"	7200000001
2		"0.001"	1801
3		"0.009"	1809
4		"0.1"	1
5	0.0 < floats < 1.0	"0.52"	137
6		"0.555"	2300
7		"0.69"	153
8		"0.9"	9
9		"0.9901"	35911
10		"0.999"	2700
	1.0		
11		"1.001"	2701
12		"1.2"	11
13		"1.8882"	43994
14	1.0 < floats < 2.0	"1.9"	18
15		"1.99"	270
16		"1.90909090901"	1071818181811
17		"1.999"	3600
	2.0		
18		"37321.9"	3935898
19		"444289.2"	48998603
20		"800445.777"	5400401200
21	2.0 < floats < 1000000	"999999.1345"	56699992211
22		"5.1"	46
23		"30.9"	1179
24		"114.7"	19033
25		"2553.1"	292978
	1000000.0		

The solution : Within a given character size the float values increment from smallest to largest. So we must order by character size first and within character sizes by inherent value sizes.

It all makes perfect sense, as a matter of fact this ordering is the only one which does.

Once the L_&_B pyramid is deployed we arrive at a "paradise" much different than Cantor's as described by David Hilbert in 1926. Here we do have order, predictability, elegance and beauty. What more do we want ?

Demonstrating the correct progression
from smallest toward larger float string character sizes.
Table 3.b

Case #	Float String character sizes	Float string bracket sizes	Float strings in order of increasing values	Integer sequence numbers
1			"0.1"	1
2	3	1	"5.1"	46
3			"9.9"	90
4			"0.01"	91
5		1	"5.01"	541
6	4		"9.99"	990
7			"10.1"	991
8		2	"55.1"	1396
9			"99.9"	1800
10			"0.001"	1801
11		1	"5.001"	6301
12			"9.999"	10800
13			"10.01"	10801
14	5	2	"55.01"	14851
15			"99.99"	18900
16			"100.1"	18901
17		3	"550.1"	22951
18			"999.9"	27000
19			"0.0001"	27001
20		1	"5.0001"	72001
21			"9.9999"	117000
22			"10.001"	117001
23		2	"55.001"	157501
24	6		"99.999"	198000
25			"100.01"	198001
26		3	"550.01"	238501
27			"999.99"	279000
28			"1000.1"	279001
29		4	"5500.1"	319501
30			"9999.9"	360000

The grand finale : Merged the two pyramids, to show the corresponding int_seqno vs. float_story values at the bracket begin/end locations. This would be the appropriate place to start praising the exceptionally utilitarian features of these two pyramids and their merged version.

The merged display showcases the duality of integer and float values over a very tiny subset of the otherwise unlimited extent of the pyramids. However, we must remember that pyramids are virtual as they take up zero storage. They serve only as a scaffold to develop the conversion algorithms. It is these algorithms which perform the quintessential twin tasks of :

Access the Level_&_Bracket pyramid to retrieve integer sequence numbers or floats

1. Compute the location of the known int_seqno in the virtual pyramid, then back-compute the float which always resides at that virtual location.
2. Compute the location of a known float in the virtual pyramid, then back-compute the int_seqno which also resides there.

Aided by the place value diagram depicted on Figure 6.3 we can device a strategy
to establish 1:1 correspondence between floats and integers

1. Count them as you count integers :
2. Count, starting from the smallest character lengths : For integers it was 1, for floats it is 3.

3. Count every float with the same character length from smallest to largest value
   just like you do for integers.

4. Value ordering for floats is accomplished by following the steps in 4.a – 4.f
   4.a start counting in the smallest level which is "level 3"
   4.b start with the leftmost decimal point bracket, and find the smallest value
       which is always of the form 0.1 or 0.01, …. 0.001 for the first bracket

   4.c within the same bracket count the floats as if they were integers; except :
       4.c.1 ignore the decimal point
       4.c.2 ignore leading zeros
       4.c.3 skip all floats with trailing zeros : e.g. 0.08, 0.09 , (skip 0.10) , 0.11
           You would sound out the numbers as follows : " eight, nine, skip 10, eleven"

   4.d shift the decimal point to the right and repeat 3 steps in step 4.c above
   4.e count the floats in all brackets of the same level repeating steps 4.c and 4.d
   4.f go to the next level, repeat steps 4.b – 4.e

This association of every positive integer with every possible float_story in existence is without doubt the number one quintessential algorithm able to retrieve a valid float.

We managed to construct a hierarchy called a "**Level_&_Bracket**" pyramid which is no longer a tree structure, but is more like a set of shelves. A set of algorithmic rules are devised which will do the two-way access. It is a work of beauty and simplicity (sort of).

Two relatively straightforward (although non- trivial) algorithms - accepting one input parameter each - will perform all the access, enforcing syntax rules declared prior and available to the users one input parameter each. The outputs are also a single quantity (additional debugging or explanatory parameters which are also returned and can be optionally made visible).

Encapsulate the two-way access formulas into Maple 2018 procedures

1. For any float_story I have provided the means to compute the int_seqno, which is the corresponding integer sequence number . The counting algorithm assigns a pre-ordained counting number to any float. The sizes of the in theory may be limitless.

This assignment will be executed by invoking a Maple-2018 procedure of the form :

int_seqno: = find_int_seqno_from_float_story (float_story)

2. For any positive integer int_seqno ranging from 1 to an unlimited number VBN there is a way to retrieve the corresponding float_story which owns this sequence number.

The action can be performed by invoking a Maple-2018 procedure of the form :

float_story: = find_float_story_from_int_seqno (int_seqno)

At this point we have at our disposal a simple and reliable method of counting floats. It is inconceivable that there will be any flaws discovered : first because of the sound mathematical foundation and second because the amazing simplicity of the algorithms.

It is important to keep in mind that each float string will conform to the precise syntax rules we initially specified above. I did drive the definition of the floats to its ultimate absolute minimalistic state which allows the creation of all possible floats but tolerates no frills or sloppiness. I omitted the [+] sign, as the absence of a sign defaults to positive floats. The syntax check are implemented and the rules are enforced in the relevant algorithms

Derive the Level_&_Bracket pyramid deposit and retrieval algorithms :

In the decimal table, X is any integer from 0-9, while Y ranges from 1-9
So when we see an X, we substitute a multiplier of 10 and for Y a multiplier of 9
What an incredible step toward a compact representation of all possible floats.
We have converted a thoroughly unmanageable tree structure into a series of formulas
as shown below. Good bye to the messy complete trees !

This would be a good time to explain the reason why the Y-range does not include 0
This follows from the two rules we postulated for floats : rule #4. Trailing zeros are prohibited.
rule #5. A leading zero is allowed only if it is immediately followed by a decimal point,
    or to put it another way when the decimal point is the second character of the float

Table 4 : Level_&_Bracket pyramid numerical details

```
================== L = 3 STRUCTURE ==============================
X.Y      = 10 * 9            =   90  =  90 * 10^(L-3), br_id = 1
================== L = 4 STRUCTURE ==============================
X.XY   = 10 * 10 *  9      =  900  =  90 * 10^(L-3), br_id = 1
YX.Y   =  9 * 10 *  9      =  810  =  81 * 10^(L-3), br_id = 2
================== L = 5 STRUCTURE ==============================
X.XXY  = 10 * 10 * 10 *  9  = 9000  =  90 * 10^(L-3), br_id = 1
YX.XY  =  9 * 10 * 10 *  9  = 8100  =  81 * 10^(L-3), br_id = 2
YXX.Y  =  9 * 10 * 10 *  9  = 8100  =  81 * 10^(L-3), br_id = 3
================== L = 6 STRUCTURE ==============================
X.XXXY = 10 *10 * 10 * 10 * 9 = 90000 =  90 * 10^(L-3), br_id = 1
YX.XXY =  9 *10 * 10 * 10 * 9 = 81000 =  81 * 10^(L-3), br_id = 2
YXX.XY =  9 *10 * 10 * 10 * 9 = 81000 =  81 * 10^(L-3), br_id = 3
YXXX.Y =  9 *10 * 10 * 10 * 9 = 81000 =  81 * 10^(L-3), br_id = 4
```

Each float_story resides on a level, then within a level in a bracket, which is accessed by a so-called bracket_id.

This is the solitary data structure which forms the backbone of the entire conversion application between an int_seqno and a float_story. The L&BP will hold the an endless number of positive floats with their associated integer sequence numbers.

Yet, incredibly enough there is not a single byte of permanent storage involved. The entire storage and retrieval scheme rests on algorithmic rules, therefore its extent is virtually unlimited.

Properties of the L_&_B pyramid

1. Each location in the L_&_B pyramid holds one int_seqno and one float_story.

2. A float_story or an int_seqno value completely determines the location they are "stored" in.

3. A pair of **Level_&_Bracket** values determine both the int_seqno and the float_story.

4. The first level is level #3, since the very first float_story equals 0.1 thus the shortest length float story is 3.

5. The level sequence number or L equals the Length of the float_story string, if we know one we know the other.

6. The number of brackets on each level is always 2 fewer than the level sequence number.

7. Thus the first level (which is level #3) has a single bracket and the fifth level is L = 7 with the br_id's ranging from 1 to 5

8. Each level has a start and finish value, so do brackets.

9. The start of any level coincides with the start of the first bracket on that level,

10. The end of the each level equals the end of the last bracket on that level

. Table 5: Decimal L_&_B pyramid

L = 3 [1, 90] size = 90
id = 1

L = 4 [91, 990], [991, 1800] size =1710
id = 1   2

L = 5 [1801, 10800], [10801, 18900], [18901, 27000] size =25200
id = 1   2   3

L = 6 [27001, 117000], [117001, 198000], [198001, 279000], [279001, 360000] size =333000
id =1   2   3   4

Example : To locate the L_&_B values for int_seqno = 230000 we advance
through the **Level_&_Bracket** compartments starting at L = 3 and moving
to the right (increased br_id) and down (increased Level). When we locate a bracket whose
starting value is less than or equal to 230000 and its terminal value is greater than or equal to
230000 we stop. The conditions we just spoke of is satisfied in the bracket with the parameters
L = 6 and br_id = 3. We can also see that the offset into this bracket is :
offset = int_seqno – br_begin = 230000 – 198001 = 31999

Table 6 : Level_&_Bracket Tables for base-10
(contents are identical ,format is different, compared to pyramid above)

LEVEL= 3

BR_ID= 1, START =   1, STORY = "0.1", END = 90, STORY = "9.9"

LEVEL= 4

BR_ID= 1, START = 91, STORY = "0.01", END = 990, STORY = "9.99"
BR_ID= 2, START = 991, STORY = "10.1", END = 1800, STORY = "99.9"

LEVEL=5

BR_ID= 1, START = 1801, STORY = "0.001", END = 10800, STORY = "9.999"
BR_ID= 2, START = 10801, STORY = "10.01", END = 18900, STORY = "99.99"
BR_ID= 3, START = 18901, STORY = "100.1", END = 27000, STORY = "999.9"

LEVEL= 6

BR_ID= 1, START = 27001, STORY = "0.0001", END = 117000, STORY = "9.9999"
BR_ID= 2, START = 117001, STORY = "10.001", END = 198000, STORY = "99.999"
BR_ID= 3, START = 198001, STORY = "100.01", END = 279000, STORY = "999.99"
BR_ID= 4, START = 279001, STORY = "1000.1", END = 360000, STORY = "9999.9"

Conclusion : Every float can be converted to its unique integer equivalent

The Level_&_Bracket pyramid appears to us as a giant data structure, which as the floats get longer and longer or the integer counters are selected to be an extremely high value then the whole edifice would become useless.

Nothing could be further from the truth. The L&B pyramid is not real, but virtual.

It is a meeting place of every valid floating string, each representing a valid floating point number and an arbitrarily large stream of corresponding integer counting or sequence numbers.
Give us an int_seqno and we deliver a valid float, or use a procedure which accepts a valid float (as a string) and it yields to us its universal integer sequence number int_seqno, without fail and without limits.

There is absolutely no need to sift through an impossibly and ever growing junk pile of candidate floats containing both gibberish and valid instances of floats.. The L&B pyramid has only bona-fide perfectly constructed floats.

Although the two-way access algorithms are not trivial, they are elementary, easily designed and implemented in two Maple1-18 procedures. A single call to either will produce the desired floats or integer sequence numbers.

The procedures will traverse through the entire virtual L&B data structure in a few steps whose number increments with the string length of the floats or with the magnitude of the integer sequence numbers, but it is a logarithmic increase instead of an exponential one.

So reasonable length float stories will be handled instantaneously. If we can specify it we can process it, subject to our computing resources. Of course it is possible to specify floats of extreme length using stacked exponentials, for example : $0.347 * (10^{13^{29^{3^{4^5}}}})$ and so on which remain out of our reach.

The ink needed to visualize such expansions would fill endless universes but it is not a reflection on the pair of algorithms but on the magnitude of the parameters.

In theory the correspondence between valid floats and integer sequence numbers is without limits.

With the algorithms developed we will be ready to have mastery of several operations whose main actors are the twin algorithms which manipulate floats and their corresponding integer sequence numbers.

List of the 6 L_&_B fundamental access operations

1. Specify a (Level, Bracket) integer pair and retrieve the corresponding ISQN value
2. Specify a (Level, Bracket) integer pair and retrieve the corresponding float.

3. Specify an arbitrary ISQN value and retrieve the corresponding (Level, Bracket) integer pair
4. Specify an arbitrary float and retrieve the corresponding (Level, Bracket) integer pair.

We may also combine the following pairs :

#5. Specify an arbitrary ISQN value and retrieve the corresponding float.
   We do this by executing operations #3 and #2 in that order.
   When we are counting floats we always execute operation #5

#6. Specify an arbitrary float value and retrieve the corresponding ISQN value.
   We do this by executing #4 and #1 in that order.

Both operations #5 and #6 are encapsulated in their respective Maple-2018 algorithms.

List of high-level counting and conversion operations
(use the fundamental operations above)

Counting : Is the act of listing all the floats corresponding to a range of integer sequence numbers (from now referenced as int_seqno) between and including a minimum and maximum int_seqno.
It is usually done in a loop.

Ordering : Line up all the floats in a set in increasing order of int_seqno's
This can be done by retrieving the int_seqno's for each float in the set, then sorting the
pair of integers and floats in order of increasing int_seqno_values.

Enumeration : Obtain the number of floats between a minimum and maximum int_seqno,
or obtain the number of floats between two floats, which equals the int_seqno difference
between two floats less 1. (e.g. the number of floats between 3.4 and 8.3 is 75-31-1 = 43)

Retrieval : The ability to obtain any float by using its universal int_seqno, either
individually or in a range. This means that we can ascertain the precedence of any two floats
(which comes first in the counting sequence)

Tagging : Assign a unique int_seqno to an individual float.

Notes : All algorithms which produce valid floats can be assigned a set of integer
sequence numbers corresponding to the list of the approximate values of the algorithms.
For example take the 5 billion long π approximation sequence (the latest)
We perform all operations listed above using only a pair one-to-one algorithms.
Each takes a single argument : either a float or an int_seqno and returns its counterpart.

These two procedures demonstrate the amazing discovery : Every single float has its own unique,
easily computable positive integer counting value, or integer sequence number (the ISQN).

This demonstrated fact invalidates all attempted proofs and conclusions whose main themes are that the
floats are uncountable and that they have a higher order of cardinality than the positive integer numbers.

Footnotes :

1. It is the set of syntax rules which drives the logic of the algorithms

 The conversion algorithms which pairs a particular float string with an integer and vice-versa are somewhat counterintuitive. The algorithms although elementary are not trivial. They involves multiple steps which one will find tedious and somewhat labor intensive but easily manageable thanks to the very capable Maple-2018 mathematical software. One source of complexity is the consequence of the strict syntax rules laid down for the floats. These rules complicate the logic of the algorithms and for a particular implementation they are considered set in stone. They are as utilitarian as logically possible.

2. Counting the integers too or hybrid counting

Anybody may change the syntax rules I defined for the floats, provided that they modify the algorithms accordingly. We can argue that every integer has an implied separator after the rightmost digit which is not visible. So if we discard one of the float syntax rules, we can welcome all counting integers into the realm of floats by appending the ".0" string to each integer. Then the integer counting agents would do double duty, counting not only all the floats but all the integers in this big inclusive hybrid unlimited set. I actually wrote a procedure which did just that (the integers as subjects of the counting process take a back seat to floats)

3. Negative floats

 All statements equally apply to both positive and negative values but we confined the essay to positive integer values, which are the customarily used counting numbers. We could use negative integers to count float story strings denoting negative floats or optionally we could insist on using only positive counting numbers and count both positive and negative float story strings. It is a cosmetic choice and has no bearing on the particular integer counters / float relationships.

4. Slight inconsistency in nomenclature.

I have decided to shorten the 3-word phrase "floating point numbers" to "float" However,
 due to coding idiosyncrasies which reflect my limited experience with Maple, I found it much
easier to claw my way through the somewhat challenging conversion steps. I believe I also
mentioned that we manipulated the floats in the counting algorithms as either strings or fake integers,
due to the fact that the value properties of the floats were no use to us.

At any rate, throughout the code and the demonstration scripts I ended up using the somewhat strange variable name : float_story, which I then managed to abbreviate to "float". Either way I always mean the particular character sequence consisting of the 11-member floa symbol set : [. 0 -> 9]
No signs are acknowledged (other than the implied + sign)

5. Why only positive floats are processed

I simply did not find it a productive use of my limited resources to deal with the volume of clutter which would have resulted if I decided to handle the presence of the + sign and included the set of negative floats as well. After all the incredibly important discovery of the one-to-one correspondence between floats and integers is the main act, I made every effort to deliver that in a clear, straightforward manner.

Chapter 6b

Two alternate methods yielding identical L_&_B pyramids

Alternate #1 : Construct the L_&_B pyramids using the mantissa/exponent pairs

Note: Ideally, a float to mantissa conversion should produce a valid integer for the mantissa which in addition has the special property that it does not have any trailing zeros.

To make that rule stick, it is imperative that we deal only with valid floats, which prohibit all trailing zeros and allow only one leading zero provided that it is immediately followed by the decimal point.

Then the mantissa of a float is defined to be a positive decimal integer which cannot contain leading or trailing zeros.

The exponent of a float is a signed integer with a leading negative sign only
(no leading zeros or positive signs).

Not all math packages follow this rule, but we have no choice if we are to succeed generating the purest one-to-one conversion algorithms between the floats and integers.

Examples of valid floats in addition to mantissas and exponents of those floats :

floats =   [12.34, 0.0087, 3400.5, 0.6, 0.45, 1.4]
mantissas = [1234,   87, 34005,   6,   45,   14]
exponents = [-2,   -4,   -1, -1,   -2, -1]

The mantissa-exponent organizations shown in the OFR column of the table below will morph into the Completed FVP codes as depicted in the table below.

So what are the mechanics of converting Original Float Ranges or OFR's to FVP codes ?

We can immediately note that each OFR code is made up of a decimal point or dp and a sequence of zeroes and the letters "Y". In addition the letter "Y" is interpreted as a numerical range from 1..9.

There will be a second very similar code, the FVP codes are identical with the exception of adding range code X, which decodes to 0..9.

Table 1 : Original Float Range (OFR) floats and OFR Codes

Float Lengths	Mantissa Lengths	Original Float Ranges or (OFR)	OFR Codes	Range Counts	Total # of Floats
3	1	0.1 -> 0.9	0.Y	9	
	2	1.1 -> 9.9	Y.Y	81	90
4	1	0.01 - > 0.09	0.0Y	9	
	2	0.11 - > 0.99	0.YY	81	
	3	1.01 - > 9.09	Y.0Y	81	
	3	10.1 - > 90.9	Y0.Y	81	
	3	1.11 -> 9.99	Y.YY	729	
	3	11.1 -> 99.9	YY.Y	729	1710
5	1	0.001 --> 0.009	0.00Y	9	
	2	0.011 -> 0.099	0.0YY	81	
	3	0.101 -> 0.909	0.Y0Y	81	
	3	0.111 -> 0.999	0.YYY	729	
	4	10.01 - > 90.09	Y0.0Y	81	
	4	100.1 -- > 900.9	Y00.Y	81	
	4	1.001 -> 9.009	Y.00Y	81	
	4	1.011 -> 9.099	Y.0YY	729	
	4	1.101 -> 9.909	Y.Y0Y	729	
	4	10.11 -> 90.99	Y0.YY	729	
	4	11.01 -> 99.09	YY.0Y	729	
	4	101.1 -> 909.9	Y0Y.Y	729	
	4	110.1 -> 990.9	YY0.Y	729	
	4	1.111 -> 9.999	Y.YYY	6561	
	4	11.11 -> 99.99	YY.YY	6561	
	4	111.1 -> 999.9	YYY.Y	6561	25200
Grand total of floats from lengths 3 to 5 . . . . =>					27000

The mantissa-exponent organizations shown in the OFR column of the table above will morph into the Completed FVP codes as depicted in the table below.

The mechanics of converting OFR codes to FVP codes are listed below :
1. Find any pair of matching OFR codes, those which satisfy the following criteria
1.a Have the same length.
1.b Decimal points are in the same position
1.c With the exception of one $0 \rightarrow Y$ or $Y \rightarrow 0$ mismatch, the rest of values must match 0 for 0 and Y for Y.

2. Once a matching OFR pair is found replace it with a single FVP code.
2.a. Copy all the matching columns from the OFR code to the new FVP code.
2. b For each column with a mismatch create a new FVP column with an X in it.
All column transfers must preserve the original positions (1st column to 1st column and so on)

3. Find and convert each suitable OFR code pair to create all the Stage 2 FVP codes.
4. Find any pair of matching Stage 2 FVP codes and using steps 1 through 3 convert them Stage 3 FVP codes.

5. Now we must match $0 \rightarrow 0$ or $Y \rightarrow Y$ and $X \rightarrow X$
6. Unmatched odd codes at every stage are copied unchanged to the next stage

Repeat this process until all 0's are eliminated from the FVP codes. Then we are done.

The resulting complete FVP codes now giving us the framework to produce a unique positive integer for each and every positive float which obeys the syntax rules specified in Chapter 3b.

Table 2 : Final Vertical Placement (or FVP) codes

Final Vertical Placement (or FVP) codes (Note: The range codes are : X = 0 .. 9 and Y = 1 .. 9)						
Float Length	Mantissa Length	OFR Codes	Stage 2 FVP	Stage 3 FVP	Completed FVP Codes	Bracket Sizes
3	1	0.Y	X.Y	X.Y	X.Y	90
	2	Y.Y				
4	1	0.0Y	0.XY	X.XY	X.XY	900
	2	0.YY				
	3	Y.0Y	Y.XY			
	3	Y.YY				
	3	Y0.Y	YX.Y	YX.Y	YX.Y	810
	3	YY.Y				
5	1	0.00Y	0.0XY	X.0XY	X.XXY	9000
	2	0.0YY				
	4	Y.00Y	Y.0XY			
	4	Y.0YY				
	3	0.Y0Y	X.Y0Y	X.YXY		
	4	Y.Y0Y				
	3	0.YYY	X.YYY			
	4	Y.YYY				
	4	Y0.0Y	Y0.XY	YX.XY	YX.XY	8100
	4	Y0.YY				
	4	YY.0Y	YY.XY			
	4	YY.YY				
	4	Y00.Y	Y0X.Y	YXX.Y	YXX.Y	8100
	4	Y0Y.Y				
	4	YY0.Y	YYX.Y			
	4	YYY.Y				
Grand total of floats from lengths 3 to 5 . . . . =>						27000

Table 3 Example of converting L = 5 FVP codes to Level_&_Bracket ID structures
"============================ L = 5 STRUCTURE ========================="
"X.XXY = 10 * 10 * 10 * 9 = 9000 = 90 * 10^(L-3), br_id = 1 "
"YX.XY = 9 * 10 * 10 * 9 = 8100 = 81 * 10^(L-3), br_id = 2 "
"YXX.Y = 9 * 10 * 10 * 9 = 8100 = 81 * 10^(L-3), br_id = 3 "

Important Reminders :

Please note that identical FVP codes were already obtained from the base-10 float complete tree, by manipulating and transforming its structure.

The FVP codes then were used as the marching orders from which all the **Level_&_Bracket** formulas and algorithms were developed.
The quick derivation of the FVP codes for the second time is justified on two grounds.

First, the mantissa-exponent organization was historically the algorithmic backbone of the wonderfully useful floating point numbers idea.

As you just witnessed, the derivation of the FVP codes from the underlying float structure required no introduction and manipulation of the base-10 complete tree.

Second and even more importantly, the fact that we have arrived at the identical FVP codes using a radically different second approach makes a very strong case for these codes as follows :

The FVP codes are not just the result of some clever range-of-the moment coding tricks.

I claim that they are the absolutely unique and indispensable cornerstone of finally achieving a long-abandoned effort of establishing one-to-one float-to-integer correspondence.

We all know that as matters stood up until the moment you read this, a permanent relationship between the floats and their integer counting agents was believed to be impossible.

The fantastic bonus this project gave us is not just a proof but the pudding too.

We can talk all day about the theory, but when we are able to produce integer sequence numbers for any float - which has the proper syntax - then we have arrived.

Hopefully all counterarguments are answered and satisfied !!!

Alternate #2: Construct the L_&_B pyramids using a lucky guess

I must confess that in retrospect I have gone about the discovery of the L_&_B pyramid the hard way. During the first approach in Chapter 6a, you saw me synchronizing base-11 integer complete trees with base-10 float complete trees to prove that the floats and integers do have 1-to-1 correspondence.

Next, I employed a method of reducing the float complete tree (which contained enormous number of invalid float candidates) to a structure which contained only legal floats. Then I rotated this tree to arrive at a structure which I called Final Vertical Placement (or FVP) codes.

The second approach was a lot simpler : We started out with the mantissa/exponent representation of the floats, which after some tricky manipulation did lead us to the same Final Vertical Placement (FVP) codes.

Finally, a somewhat incidental insight occurred to me. It was triggered by a careless statement I found on the internet and I quote :

"There are more floats between 0 and 1 than there are integers."

Disproving this claim was trivial, I have done it in Chapter 4a, I repeat it here :

1. Remove the "0." prefix from the float
2. Remove all trailing zeros as well.
3. Reverse the order of digits of the truncated digit sequence.

The resulting digits constitute an integer which will exhibit one-to-one correspondence with the original float. (Note we obviously exclude the limits of 0 and 1 as they are not floats)

Floats	0.1	0.01	0.0001	0.9	0.19	0.123
Integers	1	10	1000	9	91	321

Of course anyone who has been familiar with the float/integer 1-to-1 correspondence debate would immediately point out this : Yes, finding an integer mate for each float between 0 and 1 is a minor localized victory for countability. However it comes with a huge price tag as it proves a more important claim with a much wider scope :

"For each integer range (e.g. 0-1, 1-2, ... n->n+1 there are an uncountable number of floats. "
So not only we cannot finish counting for a single range but we are faced with a similar failure for an endless progression of integers. (It is like multiplying by ∞ by ω ...which brings in the infamous Aleph-0 for integers Aleph-1 for floats cardinality claim.)

It is a reasonable conclusion and one which has certainly enjoyed the near complete support of **six generations of mathematicians, professors, teachers and authors.** I would have left it at that if it wasn't for my earlier proof where I used two complete trees to demonstrate 1-to-1 correspondence. Constructing the Final Vertical Placement (or FVP) codes in Chapter 6a gave me the final clue to the simplest proof, which I present below.

1. Instead of diving in and listing all the floats between 0 and 1, list only floats with lengths of 3.
There are 9 of them between zero and one : 0.1, 0.2, 0.3, 0.4, 0.5, 0.6, 0.7, 0.8 and 0.9
There are 81 more, 9 each between 1 and 2 : 1.1, 1.2, 1.3...1.9 and 9 each between 2 and 3 all the way to between 9 and 10.

2. A very good start : finally we managed to count ALL the floats according to SOME criteria, which is the magical REPRESENTATION SIZE (or character length). That, if it has ever been done before, did not gain traction as all float-counting attempts ran into the abyss of never-ending runs.

3. We are on a roll now. Let's count all 4-character (3-digit) floats. We will have two possible brackets corresponding to the ranges below : (with range codes : X = 0 .. 9 and Y = 1 .. 9)

the first :   X.XY or 0.01 -> 9.99   the second YX.Y or 10.1 -> 99.9
Total = 10 * 10 * 9 = 900   Total = 9 * 10 * 9 = 810   (1710 four-character floats)

For a more extensive list please see the tables below :

Decimal Level by Level_&_Bracket structure chart

```
===================== L = 3 STRUCTURE ================="
br_id = 1  X.Y      = 10 * 9                  =  90 = 90 * 10^(L-3),
===================== L = 4 STRUCTURE ================="
br_id = 1  X.XY    = 10 * 10 * 9             = 900 = 90 * 10^(L-3),
br_id = 2  YX.Y    =  9 * 10 * 9             = 810 = 81 * 10^(L-3),
"==================== L = 5 STRUCTURE ================="
br_id = 1  X.XXY  = 10 * 10 * 10 * 9       = 9000 = 90 * 10^(L-3),
br_id = 2  YX.XY  =  9 * 10 * 10 * 9       = 8100 = 81 * 10^(L-3),
br_id = 3  YXX.Y  =  9 * 10 * 10 * 9       = 8100 = 81 * 10^(L-3),
"==================== L = 6 STRUCTURE ================="
br_id = 1  X.XXXY = 10 * 10 * 10 * 10 * 9 = 90000 =  90 * 10^(L-3),
br_id = 2  YX.XXY =  9 * 10 * 10 * 10 * 9 = 81000 =  81 * 10^(L-3),
br_id = 3  YXX.XY =  9 * 10 * 10 * 10 * 9 = 81000 =  81 * 10^(L-3),
br_id = 4  YXXX.Y =  9 * 10 * 10 * 10 * 9 = 81000 =  81 * 10^(L-3),
```

I must repeat over and over : The L_&_B structure already dives into algorithmic implementation details and it includes strict syntax rules for the floats as well.

However, when boneheaded claims arise in addition to doubts whether the L_&_B structure does contain the entire valid set of floats you must remember that the ultimate authority is the unassailable one-to-one correspondence between the two complete trees we have synchronized at the beginning of Chapter 6a.

Nothing else can ever override the powerful twin data structure foundation.

Chapter 7

Level and Bracket Pyramids implementation_from_MAPLE_1

A quick numerical example easily demonstrates that the complete tree generation of float candidates and the subsequent removal of invalid floats is very inefficient.

Although the valid float yield improves with the length of the floats that is not helping much. The worst feature of the wholesale generation is that it must start from the beginning every time. We need to have a way to generate floats or integer sequence numbers using a singular random access method. In this Maple-2018 worksheet we will develop the most efficient and elegant way to accomplish our goal.

$$N\_candidates := \frac{SS\left(SS^{Lmax} - 1\right)}{SS - 1} - SS^2 - SS$$

$$N\_valid := (AS - 1)(Lmax - 2)AS^{Lmax - 2}$$

$$SS := 11$$

$$AS := 10$$

$$Lmax := 5$$

$$N\_candidates := 177023$$

$$N\_valid := 27000$$

$$waste\_ratio := 6.556407407$$

STEP 1 : Hierarchy and level structure derivations

The derivations which follow are predicated on a very strict set of syntax rules which are enforced for the floats or float stories.

To avoid confusion, let's recap the 3 different floating point types :

1. Floating point numbers are a subset of reals, and comprise both positive and negative numbers. In addition they can have compact scientific, engineering or the traditional decimal notation. They have an implied positive and negative powers of base.

2. Floats are the subset of floating point numbers of an arbitrary base.
However, in this book we will use only positive, decimal floats, see Chapter 3b for rules

Float stories possess the exact float syntax, except they are declared as string types within the Maple 2018 algorithms. There is a compelling reason for switching to strings. First, we want to guard against some unexpected default conversions of the floats to scientific notation while manipulating the floats.

Even more importantly, the essential data types of the Complete Tree Structures (CTS) are symbolic character based. Thus almost all counting activities will use integer positional arithmetic anchored in the CTS structure and will ignore the negative power association of the numerical floats. Of course the float stories can be easily converted to and from genuine floats before the counting starts and after it ends.

The very first float is "0.1" and we assign the int_seqno = 1 to it.
Since all strings of the form "x.0" are excluded (stealth integers) we will only count 90
three-character floats when arriving at "9.9". This float story will have int_seqno = 90 assigned to it.

STEP 2 : Define Alphabets and AS, the Alphabet Size

We now revisit the very elegant Level_&_Bracket pyramid virtual structures
which we discovered in Chapter 6a.

This pyramid will accommodate both the int_seqno and its corresponding float_story.

The defining parameter of each structure is the number of distinct symbols
or the alphabet size (AS) which the counting numbers (int_seqno) will employ.

The value of AS usually corresponds to the number base in which all
arithmetic manipulations will be performed.

For example AS denotes a general base, AS = 10 is decimal and AS = 2 is binary.
In this worksheet we will only handle the counting of the decimal floats.

So AS = 10 and we use the decimal symbol set [0,1,2,3,4,5,6,7,8,9]
and the decimal point [.] for the separator.

STEP 3 : Decimal Level_&_Bracket or L_&_B pyramid structure

In the decimal table, X is any integer from 0-9, while Y ranges from 1-9
So when we see an X, we substitute a multiplier of 10 and for Y a multiplier of 9

That is it !!! We have converted an unwieldy and difficult to manage tree structure into
a series of formulas as shown below. Goodbye and good riddance complete trees !!!

"Decimal Level_&_Bracket or L_&_B pyramid structure chart"

```
"========================== L = 3 STRUCTURE =========================="
      "X.Y    =  10 * 9            =  90  =  90 * 10^(L-3), br_id = 1 "
"========================== L = 4 STRUCTURE =========================="
      "X.XY   =  10 * 10 * 9       =  900  =  90 * 10^(L-3), br_id = 1 "
      "YX.Y   =  9 * 10 * 9        =  810  =  81 * 10^(L-3), br_id = 2 "
"========================== L = 5 STRUCTURE =========================="
      "X.XXY  =  10 * 10 * 10 * 9   = 9000  =  90 * 10^(L-3), br_id = 1 "
      "YX.XY  =  9 * 10 * 10 * 9    = 8100  =  81 * 10^(L-3), br_id = 2 "
      "YXX.Y  =  9 * 10 * 10 * 9    = 8100  =  81 * 10^(L-3), br_id = 3 "
"========================== L = 6 STRUCTURE =========================="
      "X.XXXY =  10 * 10 * 10 * 10 * 9 = 90000 =  90 * 10^(L-3), br_id = 1 "
      "YX.XXY =  9 * 10 * 10 * 10 * 9 = 81000 =  81 * 10^(L-3), br_id = 2 "
      "YXX.XY =  9 * 10 * 10 * 10 * 9 = 81000 =  81 * 10^(L-3), br_id = 3 "
      "YXXX.Y =  9 * 10 * 10 * 10 * 9 = 81000 =  81 * 10^(L-3), br_id = 4 "
```

STEP 4 : Observe the general pattern and create the size formula for each level.

$$\text{size\_of\_dec\_L} := 90 \; 10^{L-3} + 81 \; (L-3) \; 10^{L-3}$$
$$\text{size\_of\_dec\_L} := 9 \; 10^{L-3} \; (9\,L - 17)$$

STEP 5 : Generate the beginning and end of level formulas from the level size.

This was done by the enumeration of the three expressions :
beg_of_L, end_of_L and size_of_L,   assuming L = 3 and beg_of_L (3) = 1.

Once we have produced the expressions in a the loop, the pattern has revealed itself in spades. From this point on everything became a simple shuffling of expressions.

The secret of the dual int_seqno and float_story is finally unlocked. !!!

$$\text{beg\_of\_dec\_L} := 9 \; (L-3) \; 10^{L-3} + 1$$

$$\text{size\_of\_dec\_L} := 9 \; 10^{L-3} \; (9\,L - 17)$$

$$\text{end\_of\_dec\_L} := 90 \; (L-2) \; 10^{L-3}$$

STEP 6: Verify that the right values are produced for any level = L

"L ==", 3, "== BEGIN == ", 1, "== SIZE == ", 90, "== END == ", 90, "=="

"L ==", 4, "== BEGIN == ", 91, "== SIZE == ", 1710, "== END == ", 1800, "=="

"L ==", 5, "== BEGIN == ", 1801, "== SIZE == ", 25200, "== END == ", 27000, "=="

"L ==", 6, "== BEGIN == ", 27001, "== SIZE == ", 333000, "== END == ", 360000, "=="

"L ==", 7, "== BEGIN == ", 360001, "== SIZE == ", 4140000, "== END == ", 4500000, "=="

STEP 7 : The sum of all levels up to and including level L is the same as the end of level L.

sum_of_L_sizes:=0:
for L from 3 to 7 do
  sum_of_L_sizes:=sum_of_L_sizes + 9*(9*L-17)* 10^(L-3);
end do;

L:=7; end_of_dec_L := 9*(10*L-20)* 10^(L-3):
$$L := 7$$
"L ===", 7, "end_of_dec_L", 4500000, "=== sum_of_L_sizes === ", 4500000

STEP 8 : Bracket id or br_id calculations -- The most difficult part

The formulas used for br_id = 1 and for br_id > 1 are different !!!

"====== DECIMAL === BEG, SIZE AND END OF LEVELS ======="
$$beg\_of\_dec\_L := 9 \, (L - 3) \, 10^{L-3} + 1$$
$$size\_of\_dec\_L := 9 \, 10^{L-3} \, (9 \, L - 17)$$
$$end\_of\_dec\_L := 9 \, (10 \, L - 20) \, 10^{L-3}$$

"Obtain br_size expressions by observing the STRUCTURE MATRIX"
$$br\_size\_br\_EQ\_1 := 90 \, 10^{L-3}$$
$$br\_size\_br\_GT\_1 := 81 \, 10^{L-3}$$
$$beg\_of\_dec\_br\_EQ\_1 := 9 \, (L - 3) \, 10^{L-3} + 1$$
$$end\_of\_dec\_br\_EQ\_1 := 9 \, (L + 7) \, 10^{L-3}$$

"Now we brute force it looking for the br_id dependency"
$$br\_id := 2$$
$$beg\_of\_dec\_br\_EQ\_2 := 9 \, (L + 7) \, 10^{L-3} + 1$$
$$end\_of\_dec\_br\_EQ\_2 := 9 \, (L + 7) \, 10^{L-3} + 81 \, 10^{L-3}$$
$$br\_id := 3$$
$$beg\_of\_dec\_br\_EQ\_3 := 9 \, (L + 7) \, 10^{L-3} + 81 \, 10^{L-3} + 1$$
$$end\_of\_dec\_br\_EQ\_3 := 9 \, (L + 7) \, 10^{L-3} + 162 \, 10^{L-3}$$

84

br_id := 4

beg_of_dec_br_EQ_4 := 9 (L + 7) 10^{L-3} + 162 10^{L-3} + 1

end_of_dec_br_EQ_4 := 9 (L + 7) 10^{L-3} + 243 10^{L-3}

"rearrange begin"

beg_of_dec_br_EQ_2 := 9 (L + 7) 10^{L-3} + 1

beg_of_dec_br_EQ_3 := 9 (L + 7) 10^{L-3} + 81 10^{L-3} + 1

beg_of_dec_br_EQ_4 := 9 (L + 7) 10^{L-3} + 162 10^{L-3} + 1

"and end"

end_of_dec_br_EQ_2 := 9 (L + 7) 10^{L-3} + 81 10^{L-3}

end_of_dec_br_EQ_3 := 9 (L + 7) 10^{L-3} + 162 10^{L-3}

end_of_dec_br_EQ_4 := 9 (L + 7) 10^{L-3} + 243 10^{L-3}

STEP 9: The pattern emerged, we have the explicit formulas for beg, end, size

"====== decimal === bracket_id = 1 ======="

beg_of_dec_br_EQ_1 := 9 (L − 3) 10^{L-3} + 1

end_of_dec_br_EQ_1 := 9 (L + 7) 10^{L-3}

size_of_dec_br_EQ_1 := 90 10^{L-3}

"====== decimal === bracket_id > 1 ======="

beg_of_dec_br_GT_1 := 9 (9 br_id + L − 11) 10^{L-3} + 1

end_of_dec_br_GT_1 := 9 (9 br_id − 2 + L) 10^{L-3}

size_of_dec_br_GT_1 := 81 10^{L-3}

STEP 10: Final expressions and NUMERICAL TEST for decimal bracket_id formulas

L := 7

"====== NUMERICAL TEST decimal === bracket_id = 1 ======="

beg_of_dec_br_EQ_1 := 360001

end_of_dec_br_EQ_1 := 1260000

size_of_dec_br_EQ_1 := 900000

verify_size := 900000

"====== NUMERICAL TEST decimal === bracket_id > 1 ======="

br_id := 5

beg_of_dec_br_GT_1 := 3690001

end_of_dec_br_GT_1 := 4500000

size_of_dec_br_GT_1 := 810000

verify_size := 810000

STEP 11: Show the bracket sizes throughout the pyramid

max_L := 7

"L=", 3, [1, 900000]

"L=", 4, [1, 900000, 2, 810000]

"L=", 5, [1, 900000, 2, 810000, 3, 810000]

"L=", 6, [1, 900000, 2, 810000, 3, 810000, 4, 810000]

"L=", 7, [1, 900000, 2, 810000, 3, 810000, 4, 810000, 5, 810000]

STEP 12a: Print the base 10 stories at the limits in the pyramid

"Decimal float_stories at the limit in the L_&_Br pyramid"

[3, "=", "0.1", "9.9"]

[4, "=", "0.01", "9.99", "10.1", "99.9"]

[5, "=", "0.001", "9.999", "10.01", "99.99", "100.1", "999.9"]

[6, "=", "0.0001", "9.9999", "10.001", "99.999", "100.01", "999.99", "1000.1", "9999.9"]

[7, "=", "0.00001", "9.99999", "10.0001", "99.9999", "100.001", "999.999", "1000.01", "9999.99", "10000.1", "99999.9"]

STEP 12b : Grand finale : merge the two pyramids to show the corresponding int_seqno vs. float_story values at the bracket begin/end locations.

This would be the appropriate place to start hammering in the exceptionally fabulous design outcome of these two pyramids and their merged version.

The merged display shows the duality of integer and float values over a very tiny subset of the otherwise unlimited extent of the pyramids.

We however have a shock announcement :

The pyramids take up zero storage, they serve only as a scaffold to develop the conversion algorithms.

It is these algorithms which :

1. Compute the location of the known int_seqno in the virtual pyramid, then back-compute the float_story which always resides at that virtual location.

2. Compute the location of the known float in the virtual pyramid, then back-compute the int_seqno which always resides at that virtual location.

"I", [3, "=", 1, 90]

"F", [3, "=", "0.1", "9.9"]

"I", [4, "=", 91, 990, 991, 1800]

"F", [4, "=", "0.01", "9.99", "10.1", "99.9"]

"I", [5, "=", 1801, 10800, 10801, 18900, 18901, 27000]

"F", [5, "=", "0.001", "9.999", "10.01", "99.99", "100.1", "999.9"]

"I", [6, "=", 27001, 117000, 117001, 198000, 198001, 279000, 279001, 360000]

"F", [6, "=", "0.0001", "9.9999", "10.001", "99.999", "100.01", "999.99", "1000.1", "9999.9"]

Step 13 : The final collection of beg, size, end formulas for L, BR=1 and BR > 1

$$\text{beg\_of\_dec\_L} := 9\,(L-3)\,10^{L-3} + 1$$

$$\text{size\_of\_dec\_L} := 9\,10^{L-3}\,(9\,L - 17)$$

$$\text{end\_of\_dec\_L} := 90\,(L-2)\,10^{L-3}$$

$$\text{beg\_of\_dec\_br\_EQ\_1} := 9\,(L-3)\,10^{L-3} + 1$$

$$\text{size\_of\_dec\_br\_EQ\_1} := 90\,10^{L-3}$$

$$\text{end\_of\_dec\_br\_EQ\_1} := 9\,(L+7)\,10^{L-3}$$

$$\text{beg\_of\_dec\_br\_GT\_1} := 9\,(9\,\text{br\_id} + L - 11)\,10^{L-3} + 1$$

$$\text{size\_of\_dec\_br\_GT\_1} := 81\,10^{L-3}$$

$$\text{end\_of\_dec\_br\_GT\_1} := 9\,(9\,\text{br\_id} - 2 + L)\,10^{L-3}$$

Chapter 8

Designing_One_to_one_Counting Algorithms

Disclaimer : In this chapter I am providing design details which must have been worked out before the two main conversion procedures (proc #1 & proc #2) could be implemented in Maple 2018.

However there is no time, no resources and what is even more relevant, there is no incentive to spend valuable time on providing a detailed tutorial.

For the few professionals, either from mathematics or from software backgrounds who are interested in either hosting the algorithms in Maple 2018 or converting them to some other mainstream language, there is sufficient detail to proceed.

Depending on my future commitments I could assist someone in an effort to have the algorithms made available to the public on a dedicated website.

But as much as I was motivated to develop and publish the float counting algorithms I see very little intellectual challenge in what is after all a counting of endless floating point numbers. Yes we now can count floats... and no, it is not very exciting after all.

```
restart;
libname:

lib_twin:="C:/counting_floats_PUBLISH_Maple-2018/library_twin_proc.mla":
libname:=libname,lib_twin:

with(twin_mod):
with (StringTools):
with(ListTools):
with(ImageTools):
with(MmaTranslator[Mma]):

global_short_return := false;
```

If this is false, then a list of values are returned by a procedure. Any or all of them can be selected by accessing the list. To extract either the int_seqno or the float_story values, the initial procedure calls must end with [1] to return the first item from the list.

Essential variable names for both decimal and binary :

int_seqno	int_seqno_base
float_story	float_story_base
fake_story	fake_story_base

int_seqno_diff	=	int_seqno - int_seqno_base
fake_story_diff	=	fake_story - fake_story_base

Critical Definitions :

1. L_&_Br pyramid is the abbreviation denoting the Level and Bracket pyramid we created
   L = a positive integer specifying the Level number in the pyramid.
   br_id = is a positive integer and it specifies the bracket id number in the pyramid

2. The L_and_Br pyramid is an absolutely essential virtual data structure of the entire float_story counting enterprise.

3. Each unique location in the L_&_Br pyramid has a one-to-one relationship with an integer sequence number and a floating point string, a.k.a. float_story

4. An integer sequence number (a.k.a. int_seqno or ISQN) is a positive counting agent.

5. A float_story is a string of characters made up of a previously defined alphabet. This alphabet contains a list of characters which might be used to "write" the story.

6. In this project all float stories are made up of the digits 0-9 and the decimal point.

7. A fake_story is a decimal integer fashioned by removing the decimal point from a float_story and interpreting the resulting string as a decimal integer.

For example the float_story := "32.98" gives us the fake_story of 3298 decimal

Important : We cannot subtract one story from another, only one fake_story from another so float_story_diff does not exist

8. A level L holds all the float_stories whose string length equals L.
The minimum value of L is 3, as the shortest length float we can define is 3.

For example float_story = "0.1" (Note, as we stated before 1.0 or .5 are not valid floats)

9. Brackets are numbered from 1 to the maximum possible on each Level.

10. Each bracket is an interval on a level which is the home to those float_stories which share the same decimal point position . A br_id is a unique positive integer denoting this position of the decimal point. A couple of examples of valid float stories will make this clear :

L = 3, br_id = 1 floats : 0.1, 3.2, 9.9 (note br_id > 1 is not valid)
L = 4, br_id = 1 floats : 0.01, 3.37, 9.99
L = 4, br_id = 2 floats : 1.01, 11.8, 99.9 (note br_id > 2 is not valid)

9. Each float_story resides on a level L, then within a level in a bracket denoted by a br_id. The smallest possible story within the bracket is called the float_story_base. Keep in mind that a float_story must have exactly one dp and it can be neither a leading nor a trailing dp. (can't be the first or last in the float_story_string.)

Here is how we come up with the fake_story_base value manually :

1. Mimic the float_story by matching the length and the position of the dp
2. Fill up this template with all zeros.
3. Change the rightmost 0 to a 1.
4. If the dp is in the leftmost valid position (br_id = 1) then we are done.
   Otherwise change the leftmost 0 to a 1.
5. Obtain the fake_story_base by removing the decimal point.

Example 1 :

1. Use L = 5, br_id = 2
2. Line up 5 zeros 00000, then replace the 3-rd zero (br_id+1 = 2+1 = 3) with a decimal point
3. We now have :   00.00
4. Make this a valid float by replacing both the leading and trailing zeros with 1
5. From 00.00 we obtain 10.01 This is the *float_story_base* (without the quotes)
6. Finally we obtain the fake_story_base by removing the decimal point to get : 1001

Example 2:

Base of 32.98 is 00.00 then we make it a valid float of 10.01
Base of 7.6543 is 0.0000 then we make it a valid float of 0.0001
Note, that a leading 0 is allowed if and only if it precedes
the decimal point, which is the case here !

Next, let's demonstrate the two essential procedures which enable us
to compute the various positional parameters associated with an int_seqno
within the L_&_Br pyramid structure

Important : When the float_story is known instead of the int_seqno,
we have a much simpler task of identifying the L (level) and bracket_id (br_id)
where the float_story resides. The total length of the float_story string gives us L,
and the number of digits preceding the decimal (or binary) point provides
the correct br_id for the float story.

Example 3: for "32.98" L = 5, br_id=2

Magic : How the L_&_Br pyramid facilitates the two-way conversion
between an ISQN value and a float story string or float for short.

==

Each unique location in the L_&_Br pyramid is defined by a list of 3 integers :
[Level, Bracket, Offset] or LBO triple. Also each location has a one to one relationship with :

1. an ISQN value
2. a float story

Here is how it works. We can do one of the 4 access operations :
1. Specify a Level, Bracket and an offset value, retrieve the corresponding ISQN value.
2. Specify a Level, Bracket and an offset value, return the corresponding float_story.
3. Specify an ISQN value and retrieve the Level, Bracket and offset integer values.
4. Specify a float_story and retrieve the Level, Bracket and offset integer values.

Magic happens when we combine access operations :

5. Specify an arbitrary ISQN value and retrieve the corresponding float_story value.
   We do this by executing operations #3 and #2 in that order,
6. Specify an arbitrary float_story value and retrieve the corresponding ISQN value.
   We do this by executing #4 and #1 in that order

When we are counting real story strings, we always execute operation #5
and when we inquire about a particular real story string with a given ISQN value
then we invoke operation #6. Both operations #5 and #6 are encapsulated in
two Maple-2018 algorithms, with the corresponding procedure declarations :

These two procedures and their supporting algorithms which implement and demonstrate
the amazing discovery : Every single float_story_string representing a restricted syntax
floating point value (we call it a float) has its own singular, easily computable
positive integer counting value (the int_seqno).

This demonstrated fact obliterates all discussions, all attempted proofs and postulates
whose main conclusion was - and still is - that the real numbers are uncountable or that
the real numbers have a higher order of cardinality than the positive counting integer numbers.

Even more surprising and a bit eerie is the fact that every single real number value has been "assigned"
or associated with its own unique positive integer counting number and that this association has existed
since before one single atom appeared at the dawn of the universe or existence itself.

Neither I nor anybody else had any role in creating such an association. What I did simply was to
describe an association which existed forever.

The conversion algorithm which pairs a particular real number string with an integer is admittedly
counterintuitive. The algorithm although elementary (requiring only basic arithmetic and modulo
operations) is not trivial, as it involves over a dozen distinct steps.

Some of the complexity is he direct consequence of the strict rules which comprise the float syntax.
These rules are as spartan or utilitarian as logically possible.
Both the digits (which represent conceptual counting beads or marbles)
and the decimal point are absolutely essential. The notation is irreducibly
complex : That is, remove one part and the thing falls to pieces.

The separator character which we call the decimal point or the "period"
character simply defines the starting place of both the positive and negative
"powers of base" multipliers. Without these multipliers we would be condemned to
flip counting "beads" until our eyes pop out.

With the introduction of the "powers-of-base" abstraction we came to rule the material world. It is that
single period which separates us from and lifts us above the animal kingdom.
They can count in the realm of integers but have zero ability in handling floats.
That is amazing, no?

What is also unusual that for counting purposes we had the luxury to completely ignore
the implied negative decimal powers of the floats and treat them as what we
came to call "fake integers." We managed to design and implement all counting algorithms
by remaining strictly in the positive power integer domain.

Global formula list for Level and Brackets pyramids

"Decimal or base-10 formulas"

$$\text{beg\_of\_dec\_L} := 9\,(L-3)\,10^{L-3} + 1$$

$$\text{beg\_of\_dec\_br\_EQ\_1} := 9\,(L-3)\,10^{L-3} + 1$$

$$\text{beg\_of\_dec\_br\_GT\_1} := 9\,(9\,\text{br\_id} + L - 11)\,10^{L-3} + 1$$

```
########## Procedure compute_dec_br_offset_demo ##########

compute_dec_br_offset_demo := proc (int_seqno::integer)::list;

local L, L_found,beg_of_dec_L,br_found,beg_of_br_EQ_1,beg_of_br_GT_1,
br_id,br_base_dec,offset_dec;

#First find the Level number or L
L:=3;
L_found:=false;
while L_found = false do
  beg_of_dec_L:= 9*(L-3)*10^(L-3)+1;
  print("L=",L,"beg_of_dec_L=",beg_of_dec_L);
  if beg_of_dec_L > int_seqno then
    L_found :=true;
    print("L=",L,"is too much, decrement by 1");
    L:=L-1;
    beg_of_dec_L:= 9*(L-3)*10^(L-3)+1;
  else
    L:=L+1;
  end if;
end do;

print("L=",L,"beg_of_dec_L=",beg_of_dec_L);

#Next find the bracket number or br_id

br_found:=false;
br_id:=2;

while br_found = false do
  br_base_dec := 9*(9*br_id+L-11)*10^(L-3)+1;

  if br_base_dec > int_seqno then
    br_found :=true;
    br_id:=br_id-1;
    if br_id = 1 then #must use a different formula
      br_base_dec := 9*(L-3)*10^(L-3)+1; # bracket = 1
    else
      br_base_dec := 9*(9*br_id+L-11)*10^(L-3)+1; # bracket > 1
    end if;
    print("br_id=",br_id,"br_base_dec ",br_base_dec );
    #break;
  else
    br_id:=br_id+1;
  end if;
end do; #while
```

```
offset_dec:=int_seqno-br_base_dec;

return ([L, br_id, beg_of_dec_L, br_base_dec, offset_dec]);

end proc: ########## end compute_dec_br_offset_demo ##########
```

Examples:
int_seqno:=91; compute_dec_br_offset_demo (int_seqno);
int_seqno:=11000; compute_dec_br_offset_demo (int_seqno);
int_seqno:=360000; compute_dec_br_offset_demo (int_seqno);

$$int\_seqno := 91$$
"L=", 3, "beg_of_dec_L=", 1

"L=", 4, "beg_of_dec_L=", 91

"L=", 5, "beg_of_dec_L=", 1801

"L=", 5, "is too much, decrement by 1"

"L=", 4, "beg_of_dec_L=", 91

"br_id=", 1, "br_base_dec ", 91

$$[4, 1, 91, 91, 0]$$

"============================="

$$int\_seqno := 11000$$
"L=", 3, "beg_of_dec_L=", 1

"L=", 4, "beg_of_dec_L=", 91

"L=", 5, "beg_of_dec_L=", 1801

"L=", 6, "beg_of_dec_L=", 27001

"L=", 6, "is too much, decrement by 1"

"L=", 5, "beg_of_dec_L=", 1801

"br_id=", 2, "br_base_dec ", 10801

$$[5, 2, 1801, 10801, 199]$$

"============================="

$$int\_seqno := 360000$$
"L=", 3, "beg_of_dec_L=", 1

"L=", 4, "beg_of_dec_L=", 91

"L=", 5, "beg_of_dec_L=", 1801

"L=", 6, "beg_of_dec_L=", 27001

"L=", 7, "beg_of_dec_L=", 360001

"L=", 7, "is too much, decrement by 1"

"L=", 6, "beg_of_dec_L=", 27001

"br_id=", 4, "br_base_dec ", 279001

$$[6, 4, 27001, 279001, 80999]$$

"============================="

We cannot readily say which float is first and which is the last in the counting order, which is actually the same as the order of generation : we start with "0.1" "999.9" comes later, so it is found later in the counting order

There is an easy way to figure the order of generation just by observation.
1. Check the Length of the float string : longer length => greater int_seqno
2. For same length : Check the position of dp : greater dp_pos =>greater int_seqno
3. For same Length and same dp_pos : compare the fake_int value :
   greater fake_int =>greater int_seqno
Even easier if I write a short procedure...
Easiest if I retrieve their associated int_seqno :
   greater int_seqno => float_story generated later
In the example below, there are 340874 float_stories between "125.9" and "0.00007" counting the limits
*);
> (* Demonstrating retrieval for br_id = 1 and br_id > 1
Using procedure 1 ### find_int_seqno_from_float_story
For each of the 2 procedures (1, 2) when retrieving either the float_story, or the int_seqno, there is a fork in the execution path depending on the br_id

From a known float_story string retrieve the corresponding int_seqno_base integer string using the following steps:

1. Obtain the L value by issuing a Length(float_story) function call

2. Obtain the br_id value by searching for the position of the decimal point.

Compute the int_seqno_base for the L and br_id values just obtained by selecting the appropriate from the 2 formulas listed

There are 4 formulas computing the int_seqno_base, which is the collective variable name for the various incarnations of the integer bracket beginnings.

$$\text{beg\_of\_dec\_L} := 9\,(L-3)\,10^{L-3} + 1$$

$$\text{beg\_of\_dec\_br\_EQ\_1} := 9\,(L-3)\,10^{L-3} + 1$$

$$\text{beg\_of\_dec\_br\_GT\_1} := 9\,(9\,\text{br\_id} + L - 11)\,10^{L-3} + 1$$

procedure 1 : find_int_seqno_from_float_story := proc (float_story::string)::integer;");
#Note : we verify results using procedure 2 developed after he analysis below is done...
sort of a retroactive double-check, can't hurt !!

```
print("############## br_id = 1 ##############");

L:=5; br_id:=1;

float_story:=      "3.298";
fake_story:=         3298;

float_story_base:="0.001";
fake_story_base:=        1;

int_seqno_base := 9*(L-3)*10^(L-3)+1; #br_id = 1

fake_story_diff := fake_story - fake_story_base;

fake_story_diff_DIV_by_10:=iquo(fake_story_diff,10);

print("int_seqno:=int_seqno_base + fake_story_diff -
fake_story_diff_DIV_by_10");
int_seqno:=int_seqno_base + fake_story_diff - fake_story_diff_DIV_by_10;

int_seqno_verify:=find_int_seqno_from_float_story("3.298")[1]; # using proc
1

print("############## br_id = 2 ##############");

unassign('fake_story','fake_story_base','fake_story_diff',
'int_seqno','int_seqno_base','int_seqno_diff');

L:=5; br_id:=2;

float_story:="32.98";
fake_story:=3298;

float_story_base:="10.01";
fake_story_base:=1001;

int_seqno_base:= 9*(9*br_id+L-11)* 10^(L-3)+1; #br_id > 1

fake_story_diff:=fake_story - fake_story_base;

fake_story_diff_DIV_by_10:=iquo(fake_story_diff, 10 ); #The Maple-2018 DIV
operator

print("int_seqno:=int_seqno_base + fake_story_diff -
fake_story_diff_DIV_by_10");
int_seqno:=int_seqno_base + fake_story_diff - fake_story_diff_DIV_by_10;
int_seqno_verify:=find_int_seqno_from_float_story("32.98")[1];
```

"############## br_id = 1 ##############"

$$L := 5$$

$$br\_id := 1$$

$$float\_story := \text{"3.298"}$$

$$fake\_story := 3298$$

float_story_base := "0.001"

fake_story_base := 1

int_seqno_base := 1801

fake_story_diff := 3297

fake_story_diff_DIV_by_10 := 329

"int_seqno:=int_seqno_base + fake_story_diff - fake_story_diff_DIV_by_10"

int_seqno := 4769

int_seqno_verify := 4769

"############### br_id = 2 #############"

L := 5

br_id := 2

float_story := "32.98"

fake_story := 3298

float_story_base := "10.01"

fake_story_base := 1001

int_seqno_base := 10801

fake_story_diff := 2297

fake_story_diff_DIV_by_10 := 229

"int_seqno:=int_seqno_base + fake_story_diff - fake_story_diff_DIV_by_10"

int_seqno := 12869

int_seqno_verify := 12869

Procedure 2 : find_float_story_from_int_seqno:= proc (int_seqno::integer)::string;

Note : we verify results using proc 2 developed after the analysis below is done...sort of a retroactive double-check, can't hurt !!

```
unassign('fake_story','fake_story_base','fake_story_diff',
'int_seqno','int_seqno_base','int_seqno_diff');

print("############### br_id = 1 #############");

int_seqno := 4769; #This is the only known quantity First we get L and br_id
dec_pyramid:=compute_dec_br_offset_demo (int_seqno); #procedure 4 in library
L:= dec_pyramid[1];          # L = 5  of the int_seqno in the decimal pyramid
br_id:= dec_pyramid[2];      # br_id =1 the br_id of int_seqno on level L

beg_of_dec_br_EQ_1   := 9*(L-3)*10^(L-3)+1; #br_id = 1;
int_seqno_base:=beg_of_dec_br_EQ_1;    #must be  = 1801

fake_story_base:=1; # from 0.001

int_seqno_diff := int_seqno-int_seqno_base;
int_seqno_diff_DIV_by_9 := iquo(int_seqno_diff, 9 );

print("fake_story= fake_story_base + int_seqno_diff+
int_seqno_diff_DIV_by_9");
fake_story:= fake_story_base + int_seqno_diff+ int_seqno_diff_DIV_by_9;

# to get float story, insert decimal point at br_id position and brace with
""-s
float_story:="3.298";
float_story_verify:=find_float_story_from_int_seqno(4769)[1]; # using proc 2

print("############## br_id = 2 #############");

unassign('fake_story','fake_story_base','fake_story_diff',
'int_seqno','int_seqno_base','int_seqno_diff');

int_seqno := 12869; #This is the only known quantity First we get L and
br_id
dec_pyramid:=compute_dec_br_offset_demo (int_seqno); #procedure 4 in library
L:= dec_pyramid[1];          # L = 5  of the int_seqno in the decimal pyramid
br_id:= dec_pyramid[2];      # br_id =2 the br_id of int_seqno on level L

beg_of_dec_br_GT_1   := 9*(9*br_id+L-11)* 10^(L-3)+1; #Now get the
int_seqno_base:=beg_of_dec_br_GT_1;    #must be  = 10801;

fake_story_base:=1001; # from 10.01 where L = 5 and br_id = 2

int_seqno_diff := int_seqno-int_seqno_base;
int_seqno_diff_DIV_by_9 := iquo(int_seqno_diff, 9 );

print("fake_story= fake_story_base + int_seqno_diff+
int_seqno_diff_DIV_by_9");
fake_story:= fake_story_base + int_seqno_diff+ int_seqno_diff_DIV_by_9;

# to get float story, insert decimal point at br_id position and brace with
""-s
```

```
float_story:="32.98";
float_story_verify:=find_float_story_from_int_seqno(12869)[1]; # using proc
2
```

"############### br_id = 1 ###############"

int_seqno := 4769

"L=", 3, "beg_of_dec_L=", 1

"L=", 4, "beg_of_dec_L=", 91

"L=", 5, "beg_of_dec_L=", 1801

"L=", 6, "beg_of_dec_L=", 27001

"L=", 6, "is too much, decrement by 1"

"L=", 5, "beg_of_dec_L=", 1801

"br_id=", 1, "br_base_dec ", 1801

dec_pyramid := [5, 1, 1801, 1801, 2968]

L := 5

br_id := 1

beg_of_dec_br_EQ_1 := 1801

int_seqno_base := 1801

fake_story_base := 1

int_seqno_diff := 2968

int_seqno_diff_DIV_by_9 := 329

"fake_story= fake_story_base + int_seqno_diff+ int_seqno_diff_DIV_by_9"

fake_story := 3298

float_story := "3.298"

float_story_verify := "3.298"

"############### br_id = 2 ###############"

int_seqno := 12869

"L=", 3, "beg_of_dec_L=", 1

"L=", 4, "beg_of_dec_L=", 91

"L=", 5, "beg_of_dec_L=", 1801

"L=", 6, "beg_of_dec_L=", 27001

"L=", 6, "is too much, decrement by 1"

"L=", 5, "beg_of_dec_L=", 1801

"br_id=", 2, "br_base_dec ", 10801

dec_pyramid := [5, 2, 1801, 10801, 2068]

L := 5

br_id := 2

beg_of_dec_br_GT_1 := 10801

int_seqno_base := 10801

fake_story_base := 1001

int_seqno_diff := 2068

int_seqno_diff_DIV_by_9 := 229

"fake_story= fake_story_base + int_seqno_diff+ int_seqno_diff_DIV_by_9"

fake_story := 3298

float_story := "32.98"

float_story_verify := "32.98"

Finally, here we are demonstrating the calling sequence and the return list for each of
the 2 main-line procedures. For those who would simply use the procedures to do
the two-way conversions between int_seqno and float_story this is all what they need.

```
global_short_return:=false:
global_trace_print := false:
printlevel:=0:
print("#### procedure 1  demo : find_int_seqno_from_float_story #### ");

float_story_given:="32.98";
int_seqno_expected := 12869:

results:=find_int_seqno_from_float_story (float_story_given);

int_seqno_obtained:=results[1]:
float_story_given:=results[2]:
L:=results[3];
br_id:=results[4];
fake_story:=results[5];
fake_story_base:=results[6];
fake_story_diff:=results[7];
diff_adjust:=results[8];
int_seqno_base:=results[9];

print("int_seqno_expected=",int_seqno_expected);
print("int_seqno_obtained=",int_seqno_obtained);

print("########## end of proc 1 demo ##########");
```

"########## procedure 1 demo : find_int_seqno_from_float_story #### "

float_story_given := "32.98"

results := [12869, "32.98", 5, 2, 3298, 1001, 2297, 229, 10801]

L := 5

br_id := 2

fake_story := 3298

fake_story_base := 1001

fake_story_diff := 2297

diff_adjust := 229

int_seqno_base := 10801

"int_seqno_expected=", 12869

"int_seqno_obtained=", 12869

"################## end of proc 1 demo ################"


```
print("#### proc 2  demo find_float_story_from_int_seqno #######");

int_seqno_given:=12869;
float_story_expected:="32.98":

results:=find_float_story_from_int_seqno (int_seqno_given);

float_story_obtained:=results[1]:
int_seqno_given:=results[2]:
L:=results[3];
br_id:=results[4];
int_seqno_base:=results[5];
float_story_list:=results[6];
int_seqno_diff:=results[7];
fake_story_base:=results[8];
diff_adjust:=results[9];
fake_story:=results[10];
print("float_story_expected=",float_story_expected);
print("float_story_obtained=",float_story_obtained);

print("######### end of proc 2 demo ###########");
```

"######### proc 2 demo find_float_story_from_int_seqno ###########"

int_seqno_given := 12869

results := ["32.98", 12869, 5, 2, 10801, [8, 9, ".", 2, 3], 2068, 1001, 229, 3298]

L := 5

br_id := 2

int_seqno_base := 10801

float_story_list := [8, 9, ".", 2, 3]

int_seqno_diff := 2068

fake_story_base := 1001

diff_adjust := 229

fake_story := 3298

"float_story_expected=", "32.98"

"float_story_obtained=", "32.98"

"################## end of proc 2 demo ######################"

Chapter 9 MAPLE script

Library_creation_of_twin_counting_algorithms
(from MAPLE_3_Library_creation_of_twin_counting_algorithms)

```
(*
This library contains the 4 essential Maple 2018 procedures
which enable the users to execute either of the two-way
[float_story <--> int_seqno] conversions

ROUNDTRIP verification passed on a 50,000+ digit approximation of π

These procedures are the crown jewels of the "COUNTING THE FLOATS" project

You are witnessing set theory history being made right here. ENJOY !!!
*)

(*
1. To create this library, first create the directory :
   C:/counting_floats_PUBLISH_Maple-2018
2. Open a MAPLE worksheet, copy the statements between
   ###### START OF LIBRARY CREATION and ###### end OF LIBRARY CREATION
3. Execute the worksheet
4. Your library should be in the directory with the name as shown:
   C:/counting_floats_PUBLISH_Maple-2018/library_twin_proc

5. To access this library from any Maple 2018 worksheet, insert
   the following block of 9 statements at the top of the worksheet :

   Let's assume that your directory for both the library (this file ) and
   for your worksheets is : C:/counting_floats_PUBLISH_Maple-2018

restart;
libname:
lib_twin:="C:/counting_floats_PUBLISH_Maple-2018/library_twin_proc.mla":
libname:=libname,lib_twin:
with(twin_mod):

with (StringTools):
with (ListTools):
with (MmaTranslator[Mma]):

global_short_return:=true; # only first item in return list is accessed
*);
```

```
################### START OF LIBRARY CREATION ####################

lib:="C:/counting_floats_PUBLISH_Maple-2018/library_twin_proc.mla";
                lib := "C:/counting_floats_PUBLISH_Maple_2018/library_twin_proc.mla"
with (StringTools):
with (ListTools):
with (MmaTranslator[Mma]):

global_short_return:=true; # only first item in return list is accessed
                global_short_return := true

twin_mod := module()

export

find_int_seqno_from_float_story, #1 (was  7) converts float to int_seqno
Calls  #3,  Called by user
find_float_story_from_int_seqno, #2 (was  8) converts int_seqno to float
Calls  #4,  Called by user
check_for_valid_float_story,     #3 (was  9) checks syntax of floats
Calls none, Called by #1
compute_L_BR_offset_decimal;     #4 (was 10) computes L_& Br id's
Calls none, Called by #2

option package;
```

```
######### Procedure 1 find_int_seqno_from_float_story #########

find_int_seqno_from_float_story := proc (float_story::string)::integer;

local float_story_list, float_story_no_dp, float_story_base,
base_story_no_dp,
fake_story_base, fake_story_diff, diff_adjust,
int_seqno,int_seqno_base,i,RNSL,
char_pos,br_id, fake_story;

if global_trace_print = true then
  print("IN 7 find_int_seqno_from_float_story");
end if;

if check_for_valid_float_story (float_story) = false then
  return(["INVALID float_story string",float_story]);
end if;

#verifying decimal story syntax

float_story_list:=Explode(float_story); ###

for i from 1 to nops(float_story_list) do # make sure that it is decimal
  if (float_story_list[i] <> ".") and ((float_story_list[i] < "0") or
     (float_story_list[i] > "9")) then
    print (float_story, "is not a valid decimal story");
    return("FAIL");
  end if;
end do;

RNSL:= nops(float_story_list); # the number of chars in the story, including
dp !
for char_pos from 1 to RNSL do
  if float_story_list[char_pos] = "." then # we have our decimal point
    br_id:=char_pos-1;  # if float_story is [0.1234] then br id = 2-1 = 1
  end if;
end do; # br id better be between 1 and RNSL-2

#Now we have both the RNSL and the br_id values

if RNSL > 3 and br_id > 1 then # the first digit is one
  float_story_base:=[1]; # this is one of two times we can do direct
assignments
else
  float_story_base:=[0]; # this is the other one of two times we can do
direct assignments
end if;

for i from 1 to RNSL-2 do
  if i = br_id then
    float_story_base:=[op(float_story_base),"."]; #!!!!!!!!!!!!!!!!!!!! APP
  else
    float_story_base:=[op(float_story_base),0]; # we get RNSL zeros for
RNSL=3
  end if;
end do;
```

```
float_story_base:=[op(float_story_base),1]; #last digit in the float base is
always 1

fake_story_diff:=FromDigits(float_story_list)-FromDigits(float_story_base);

base_story_no_dp:=[]; #a LIST !!! strip decimal point from base list
for i from 1 to nops(float_story_base) do
  if float_story_base[i] <> "." then
    base_story_no_dp:=[op(base_story_no_dp),float_story_base[i]];
  end if;
end do;
fake_story_base:= FromDigits(base_story_no_dp); # the decimal of the base

float_story_no_dp:=[]; #strip decimal point from float_story_list

for i from 1 to nops(float_story_list) do
  if float_story_list[i] <> "." then
    float_story_no_dp:=[op(float_story_no_dp),float_story_list[i]];
  end if;
end do;

###################### MAIN ACT ############################

fake_story:= FromDigits(float_story_no_dp); # the "fake" of the incoming
float_story
fake_story_diff := fake_story - fake_story_base;

if br_id > 1 then
  int_seqno_base := 9*(9*br_id+RNSL-11)*10^(RNSL-3)+1;
else
  int_seqno_base := 9*(RNSL-3)*10^(RNSL-3)+1; #br_id = 1
end if;

diff_adjust:=iquo(fake_story_diff,10);

int_seqno:=int_seqno_base + fake_story_diff - diff_adjust;

####################################################################

if global_trace_exit= true then
  print("EXIT 7 ========= find_int_seqno_from_float_story");
end if;

if global_short_return = true then
  return (int_seqno);
else
  return ([int_seqno, float_story, RNSL, br_id, fake_story,
    fake_story_base, fake_story_diff, diff_adjust, int_seqno_base]);
end if;

end proc: ########## End 1 find_int_seqno_from_float_story ##########
```

```
############# Procedure 2 find_float_story_from_int_seqno #############

# (Note : Procedure 8 in main library )
(*
Works now, with a legitimate algorithm, but by no means trivial
on the account of the various conversions and float_story_base
management. Returns a true string containing a valid float e.g.
"2.71" after a call like below has been made :
float_story:= find_float_story_from_int_seqno(334);
To display or use the actual float number value, issue the call :

    float_num:=FromDigits(float_story);
*);

find_float_story_from_int_seqno:= proc (int_seqno::integer)::string;

local i, t, br_id, RNSL, return_list, int_seqno_base, int_seqno_diff,
diff_adjust, int_seqno_string, fake_story_list, fake_index,
template_index, fake_story, fake_story_base, float_story,template_list,
template_size, float_story_list_out,story_list_out_len;

unassign('base'); # base is used in convert as a keyword !!!!!!!!!!!!

if global_trace_print = true then
  print("IN 8 find_float_story_from_int_seqno(int_seqno)");
end if;

# make sure that int_seqno is a positive decimal number
if int_seqno <=0 then
  print (int_seqno, "is not a nonzero positive decimal number");
  return("FAIL");
end if;

return_list:=compute_L_BR_offset_decimal (int_seqno): #procedure 10b, does
everything

# return list : [RNSL=1,L_beg_dec=2,br_id=3,br_base_int=4, offset_dec=5]
RNSL:=        return_list[1];
int_seqno_base:= return_list[4];
br_id:=       return_list[3];

##### # Appending the list will increase the possible list sizes

### step 1
if RNSL > 3 and br_id > 1 then
  template_list:=[1];
else
  template_list:=[0];
end if;

### step 2
for i from 1 to br_id-1 do  # will not execute for br_id=1 I think ????
  template_list:=[op(template_list),0];
end do;

### step 3
template_list:=[op(template_list),"."]; #append the dp
```

```
### step 4
for i from 1 to RNSL-br_id-2 do  # fill with 0's from dp to next to last
char
   template_list:=[op(template_list),0];
end do;

### step 5
template_list:=[op(template_list),1]; # last digit is 1, then ranges from
1..9

#convert the base_float_story to a fake integer by removing the dp.
fake_story_base:=[]; #strip decimal point from template_list
for i from 1 to nops(template_list) do
   if template_list[i] <> "." then
     fake_story_base:=[op(fake_story_base),template_list[i]];
   end if;
end do;
fake_story_base:= FromDigits(fake_story_base); # the "fake" float story

################# MAIN ACT #################

int_seqno_diff:=int_seqno - int_seqno_base;

diff_adjust:=iquo(int_seqno_diff,9);

fake_story:= fake_story_base + int_seqno_diff + diff_adjust;

##############################################################

# We got a fake integer version of the float_story string

fake_story_list:=Reverse(convert (fake_story, base, 10));
# turn it into a list ,convert fake decimal to a float_story,
# matching the template_list of the current bracket
#let's create a float_story_list_out into which we merge template_list and
fake_story_list
#None of the lists can be written into directly. EVERY LIST MUST BE APPENDED
#we are converting a list without dp to the complete float_story_list_out

float_story_list_out:=[];

template_index:=nops(template_list); # the skeleton list with 0,1 and dp
e.g. [1,0,".",1]
fake_index:=nops(fake_story_list); # this gets merged in with the template

while (template_index > 0 ) do # we sweep the template and use its contents
   if template_list[template_index] <> "." then # it is a 0, 1 and dp
skeleton
      if ( fake_index > 0 ) then
        float_story_list_out:=
[op(float_story_list_out),fake_story_list[fake_index]];
        fake_index:=fake_index-1;
      else
        float_story_list_out:= [op(float_story_list_out),0];
      end if;
   else # just append the decimal point
```

106

```
      float_story_list_out:= [op(float_story_list_out),"."];
   end if;
   template_index:=template_index-1;
end do;  # while

story_list_out_len:= nops(float_story_list_out);

#Now we have a proper float_story in a list with dp but in reverse order !!!
float_story:="":

printlevel:=1;
for i from nops(float_story_list_out) by -1 to 1 do
   # take the list of float digits and convert into a string in reverse
order;
   float_story:=cat(float_story, float_story_list_out[i]):
end do:

if global_trace_exit = true then
   print("EXIT 8 ========== find_float_story_from_int_seqno");
end if;

if global_short_return = true then
   return (float_story);
else
   return ([float_story,int_seqno,RNSL, br_id,
int_seqno_base,float_story_list_out,
   int_seqno_diff,fake_story_base, diff_adjust, fake_story]);

end if;

end: ########## end 2 find_float_story_from_int_seqno ########
```

```
########## Procedure 3 check_for_valid_float_story ##########

check_for_valid_float_story := proc (float_story::string)::constant;

local num_of_dec_points,i, float_story_len, dec_point_char;

float_story_len:=Length(float_story);
dec_point_char := ".";
num_of_dec_points := 0;

######## ==== 0 ==== ######## story chars must be one of ".0123456789"

for i from 1 to float_story_len do
  if (float_story[i] <> ".") and (float_story[i] <"0") or
    (float_story[i] > "9") then
    print("float characters must be one of [ .0,1,2,3,4,5,6,7,8,9 ]");

    return(false);
  end if;
end do;


######## ==== 1 ==== ######## float_story length must be >= 3

if float_story_len < 3 then # we only check for stories with length >= 3
  print("float character lengths must be at least 3 ");
  return(false);
end if;


######## ==== 2 ==== ######## allow exactly one decimal point
for i from 1 to float_story_len do
  if float_story[i] = dec_point_char then
    num_of_dec_points :=num_of_dec_points+1;
  end if;
end do;

if (num_of_dec_points <> 1 ) then
  print("floats must have exactly one decimal point ");
  return (false);
end if;


######## ==== 3 ==== ######## do not allow leading decimal points
if float_story[1] = dec_point_char then
  print("floats do not allow leading decimal points");
  return(false);
end if;


######## ==== 4 ==== ######## do not allow trailing decimal points
if float_story[float_story_len] = dec_point_char then
  print("floats do not allow trailing decimal points");
  return(false);
end if;


######## ==== 5 ==== ######## leading 0 must be followed by a dp.
if float_story[1] = "0" and float_story[2] <> "." then
  print("floats leading 0 must be immediately followed by a decimal point");
  return(false);
end if;
```

```
######## ==== 6 ==== ######## reject trailing zeros
if ( float_story[float_story_len] = "0") then
  print("floats cannot have trailing 0 digits ");
  return (false);
end if;

return (true);

end proc: ########## End 3 check_for_valid_float_story ##########
```

```
########## Procedure 4 compute_L_BR_offset_decimal ##########

compute_L_BR_offset_decimal := proc (int_seqno::integer)::list;

local L, level_found,L_beg_dec,bracket_found,beg_of_AS_br_EQ_1,
beg_of_AS_br_GT_1,br_id,br_base_dec,offset_dec;

L:=3;
level_found:=false;
while level_found = false do
  L_beg_dec:= 9*(L-3) * 10^(L-3)+1; #
  if L_beg_dec > int_seqno then
    level_found :=true;
    L:=L-1;
    L_beg_dec:= 9*(L-3) * 10^(L-3)+1; #
  else
    L:=L+1;
  end if;
end do;

bracket_found:=false;
br_id:=2;

while bracket_found = false do
  beg_of_AS_br_GT_1 := 9*(9*br_id+L-11)*10^(L-3)+1;
  if beg_of_AS_br_GT_1 > int_seqno then
    bracket_found :=true;
    br_id:=br_id-1;
    if br_id = 1 then #must use a different formula
      br_base_dec := 9*(L-3) * 10^(L-3)+1; #bracket = 1
    else
      br_base_dec := 9*(9*br_id+L-11)*10^(L-3)+1; #bracket > 1
    end if;
  else
    br_id:=br_id+1;
  end if;
end do; #while

offset_dec:=int_seqno-br_base_dec;
return ([L, L_beg_dec, br_id, br_base_dec, offset_dec]);
end proc: ########## end 4 compute_L_BR_offset_decimal ##########
end module:

savelib('twin_mod', lib):

with (twin_mod ):

##### END OF LIBRARY CREATION
############### TEST LIBRARY CREATION ##########
```

```
# Verifying the syntax check logic of float-like string inputs
compute_L_BR_offset_decimal (3);

int_seqno1:=find_int_seqno_from_float_story ("-1.123"):
int_seqno1:=find_int_seqno_from_float_story ("1."):
int_seqno1:=find_int_seqno_from_float_story ("1123."):
int_seqno1:=find_int_seqno_from_float_story (".1123"):
int_seqno1:=find_int_seqno_from_float_story ("1.1.23"):
int_seqno1:=find_int_seqno_from_float_story ("001.123"):
int_seqno1:=find_int_seqno_from_float_story ("1.1230"):
```

$$[3, 1, 1, 1, 2]$$

"float characters must be one of [. 0, 1, 2, 3, 4, 5, 6, 7, 8, 9]"

"float character lengths must be at least 3 "

"floats do not allow trailing decimal points"

"floats do not allow leading decimal points"

"floats must have exactly one decimal point "

"floats leading 0 must be immediately followed by a decimal point"

"floats cannot have trailing 0 digits "

```
global_short_return:=true; # only first item in return list is accessed

float_story_in:="1.234554321";
int_seqno:=find_int_seqno_from_float_story (float_story_in);
 # only int_seqno is returned
float_story_out:= find_float_story_from_int_seqno (int_seqno);
int_seqno_check:=find_int_seqno_from_float_story (float_story_out);
 # only float_story is returned

if (int_seqno_check = int_seqno ) then
  print("ROUNDTRIP IS VERIFIED !!!");
end if;
```

global_short_return := true

float_story_in := "1.234554321"

int_seqno := 8311098889

float_story_out := "1.234554321"

int_seqno_check := 8311098889

"ROUNDTRIP IS VERIFIED !!!"

```
global_short_return:=false; # To access all items on the return list,
  must specify indices
float_story_in:="12.34";
int_seqno:=find_int_seqno_from_float_story (float_story_in)[1];
float_story_out:= find_float_story_from_int_seqno (int_seqno)[1];
if (float_story_out = float_story_in ) then print("ROUNDTRIP IS VERIFIED");
end if;
```

global_short_return := false

float_story_in := "12.34"

111

int_seqno := 11011

float_story_out := "12.34"

"ROUNDTRIP IS VERIFIED !!!"

```
float_story_in:="1.234554321";
int_seqno_out:=find_int_seqno_from_float_story (float_story_in);
float_story_out:= find_float_story_from_int_seqno (int_seqno_out[1]);
# retrieve int_seqno from list

params:=compute_L_BR_offset_decimal (32133);
print("returned parameters :
    [Level, Level_begin_decimal, bracket_id, bracket_base_decimal,
offset_decimal]");
```

float_story_in := "1.234554321"

int_seqno_out := [8311098889, "1.234554321", 11, 1, 1234554321, 1,
       1234554320, 123455432, 7200000001]

float_story_out := ["1.234554321", 8311098889, 11, 1, 7200000001, [1, 2, 3,
      4, 5, 5, 4, 3, 2, ".", 1], 1111098888, 1, 123455432, 1234554321]

params := [6, 27001, 1, 27001, 5132]

"returned parameters : [Level, Level_begin_decimal, bracket_id,
      bracket_base_decimal, offset_decimal]"

Chapter 10 MAPLE script

Float outputs of twin counting algorithms (PI, 1/3, 1/7, etc)
(from MAPLE_4_Float_outputs_of_twin noteworthy_algorithms)

```
restart;
libname:

twin_proc:="C:/counting_floats_PUBLISH_Maple-2018/library_twin_proc.mla":
libname:=libname,twin_proc;

with(twin_mod):
           libname:="C:\Program Files\Maple 18\lib", ".",
               "C:/counting_floats_PUBLISH_Maple_2018/library_twin_proc.mla"

with(StringTools):
with(ListTools):
#with(ImageTools):
with(MmaTranslator[Mma]):
with(combinat):
with(SumTools):


#### ATTENTION ####

global_execute:=true; # if false, will not execute each and every case study
global_short_return :=false;

# set this to true to see only int_seqno or float_story for procedures 7&8
float_story1:=find_float_story_from_int_seqno (90); #returns the entire list
float_story1:=find_float_story_from_int_seqno (90)[1]; #returns float story
only
```

$$global\_execute := true$$

$$global\_short\_return := false$$

$$float\_story1 := ["9.9", 90, 3, 1, 1, [9, ".", 9], 89, 1, 9, 99]$$

$$float\_story1 := "9.9"$$

```
(* !!!!! SYNTAX RULES WHICH APPLY TO FLOAT STRINGS !!!!!
1. Only the 11 symbols [ . , 0, 1, 2, 3, 4, 5, 6, 7, 8, 9 ] may be used.
2. Use the decimal point [ . ]  exactly once.
3. No trailing or leading decimal points are permitted.
4. Trailing zeros are prohibited.
5. A single leading zero is allowed only if it is followed by a decimal
point

Note: Features which were implemented in the project but not included here
1. Scientific notation can be converted to float
2. We can handle negative floats
3. I did implement any-base AS conversions (both have to be in base AS )
4. Did do hybrid sequences (where "x.0" are allowed as valid floats )
4. I once messed around hybrid floats (where integer.0 is counted as a
float)
5. Even mantissa/exponent representation of floats could be converted to
floats
6. Outputs of algorithms can be counted if converted to floats
```

Sample outputs of the two flagship conversion procedures

\# 1 ######### Return both strings and valid floats from int_seqno

\# 2 ######### Number of floats between and including two floats

\# 3 ######### Floats for Fibonacci_integers

\# 4 ######### Generate float_story by adding to int_seqno

\# 5 ######### ROUNDTRIP of the two essential conversion procedures

\# 6 ######### Generate a single int_seqno's from a single fraction

\# 7 ######### Int_seqno-1 causes a huge jump in float story value

\# 8 ######### Find float stories for powers of 10 int_seqno's

\# 9 ######### Counting π approximations

\# 10 ######### No floats can be inserted between two adjacent floats

\# 11 ######### Two-way conversion of very large integers or floats

\# 12 ######### A repetitive pattern, in spite of the modulo arithmetic

\# 13 ######### Generating repeating int_seqno from repeating floats

\# 14 ######### List of odd patterns of converted repeating decimals

\# 15 ######### Strange coincidence : int_seqnos equal fake float_stories

\# 16 ######### How to find the number of floats between any two floats

\# 17 ######### Integer output of Shifting decimal points to right

```
# 1 ######### Return both strings and valid floats from int_seqno

int_seqno:=13062:
all_parameters:=find_float_story_from_int_seqno (int_seqno);
float_string:=all_parameters[1];
list_story:=all_parameters[6];
myfloat:=FromDigits(list_story); # a genuine float

         all_parameters :=["35.13", 13062, 5, 2, 10801, [3, 1, ".", 5, 3],
            2261, 1001, 251, 3513]

                         float_string := "35.13"
                      list_story := [3, 1, ".", 5, 3]
                            myfloat := 31.53

# 2 ######### Number of floats between and including two floats

float_story_1:="1.5":
float_story_2:="0.2":
 int_seqno_1:= find_int_seqno_from_float_story(float_story_1)[1]:
 int_seqno_2:= find_int_seqno_from_float_story(float_story_2)[1]:
if int_seqno_2 > int_seqno_1 then
   num_fl_btw_and_incl:= int_seqno_2 - int_seqno_1+1;
else
  num_fl_btw_and_incl:= int_seqno_1 - int_seqno_2 +1;
end if:
print("number_of_floats_between",float_story_1,"and",
float_story_2,"is",num_fl_btw_and_incl);

        "number_of_floats_between", "1.5", "and", "0.2", "is", 13


# 3 ######### Floats for Fibonacci_integers

printlevel:=0:
Fibonacci_int:= [1, 2, 3, 5, 8, 13, 21, 34, 55, 89, 144, 233, 377, 610, 987,
1597, 2584, 4181, 6765, 10946, 17711, 28657, 46368, 75025, 121393, 196418,
317811]:

for n from 1 to nops(Fibonacci_int) -24 do
   float_story:=find_float_story_from_int_seqno (Fibonacci_int[n])[1]:
   print("n=",n, "Fibonacci(n)",
   Fibonacci_int[n],"float_story",float_story);
end do;
                  "n=", 1, "Fibonacci(n)", 1, "float_story", "0.1"

                  "n=", 2, "Fibonacci(n)", 2, "float_story", "0.2"

                  "n=", 3, "Fibonacci(n)", 3, "float_story", "0.3"

                  "n=", 4, "Fibonacci(n)", 5, "float_story", "0.5"
```

```
# 4 ######### Generate float_story by adding to int_seqno

printlevel:=0:
start_int:=1;
add_value:=7300000;
int_seqno:=start_int:
for i from 1 to 10 do
  float_story:=find_float_story_from_int_seqno(int_seqno)[1]:
  int_seqno_verify:= find_int_seqno_from_float_story(float_story)[1]:
  if int_seqno <> int_seqno_verify then
    print ("mismatch", int_seqno, int_seqno_verify);
  end if;
  if i > 1 then print(int_seqno, float_story); end if;
  int_seqno:=int_seqno+add_value;
end do;
```

$$start\_int := 1$$

$$add\_value := 7300000$$

$$7300001, \text{"3.111112"}$$

$$14600001, \text{"22.22223"}$$

$$21900001, \text{"133.3334"}$$

$$29200001, \text{"944.4445"}$$

$$36500001, \text{"8555.556"}$$

$$43800001, \text{"76666.67"}$$

$$51100001, \text{"677777.8"}$$

$$58400001, \text{"0.4888889"}$$

$$65700001, \text{"1.3000001"}$$

```
# 5 ######### ROUNDTRIP of the two essential conversion procedures

for i from 1 to 3 do # roundtrip
int_seqno:= i*137943;
story:=find_float_story_from_int_seqno (int_seqno)[1]:
int_seqno_back:=find_int_seqno_from_float_story (story)[1];
print(int_seqno,story,int_seqno_back);
end do;
```

$$137943, \text{"33.269"}, 137943$$

$$275886, \text{"965.39"}, 275886$$

$$413829, \text{"0.59809"}, 413829$$

```
# 6 ######### Generate a single int_seqno from a single fraction

cur_float_str:="0.0422535211267605633802816901408";
int_seqno:= find_int_seqno_from_float_story(cur_float_str)[1];
```

$$cur\_float\_str := \text{"0.0422535211267605633802816901408"}$$

$$int\_seqno := 27038028169014084507042253521268$$

```
# 7 ######## int_seqno-1 causes a huge jump in float_story value

min_int_seqno:=7200000001;
float_story1:=find_float_story_from_int_seqno(min_int_seqno)[1];
min_int_seqno:=min_int_seqno-1;
float_story2:=find_float_story_from_int_seqno(min_int_seqno-1)[1];
```

$$\text{min\_int\_seqno} := 7200000001$$

$$\text{float\_story1} := \text{"0.000000001"}$$

$$\text{min\_int\_seqno} := 7200000000$$

$$\text{float\_story2} := \text{"99999999.8"}$$

```
# 8 ######### Find float stories for powers of 10 int_seqno's

for i from 1 to 6 do
float_story:=find_float_story_from_int_seqno(10^(i))[1]:
print(10^(i),float_story);
end do;
```

$$10, \text{"1.1"}$$

$$100, \text{"0.11"}$$

$$1000, \text{"11.1"}$$

$$10000, \text{"9.111"}$$

$$100000, \text{"8.1111"}$$

$$1000000, \text{"7.11111"}$$

```
# 9 ######### Counting π approximations

shortPI_in:="3.14159265358979323846264338327950288841971693993751";
printlevel:=0: #
PI_approx:="3.14":
print("Counting PI approximations");
for i from 1 to 10 do
first_integer:= find_int_seqno_from_float_story (PI_approx)[1];
print(i, PI_approx,first_integer);
PI_approx:=cat(PI_approx, shortPI_in[i+4]):
end do;
PI_approx_zero_ending:="3.14159265358979323846264338327950";
print("Floats ending in zero cannot be valid PI approximations !!!");
int_seqno:= find_int_seqno_from_float_story (PI_approx_zero_ending)[1];
```

$$\text{shortPI\_in} :=$$
$$\text{"3.14159265358979323846264338327950288841971693993751"}$$

$$\text{"Counting PI approximations"}$$

$$1, \text{"3.14"}, 373$$

$$2, \text{"3.141"}, 4627$$

$$3, \text{"3.1415"}, 55274$$

$$4, \text{"3.14159"}, 642744$$

$$5, \text{"3.141592"}, 7327433$$

$$6, \text{"3.1415926"}, 82274334$$

```
          7, "3.14159265", 912743339

          8, "3.141592653", 10027433388

          9, "3.1415926535", 109274333882

         10, "3.14159265358", 1182743338823

                         ""

PI_approx_zero_ending:="3.141592653589793238462643383279550"

"Floats ending in zero cannot be valid PI approximations !!!"

    "floats cannot have trailing 0 digits "

    int_seqno:="INVALID float_story string"
```

10 ######## No floats can be inserted between two adjacent floats

```
(*There is this supposedly "knockout argument against countability.
They take two floats which differ only by their last digits, for example
"0.333333" and "0.333334"
then they claim that there is an infinite number of other floats can be
squeezed in. Not on my watch !
```

The whole century-old debate about float countability originated from the
realization that floats cannot be ordered or (enumerated) using their
inherent values, which has given us unlimited resolution in the physical
world. Then everybody gave up on counting the floats and subsequently
declared them uncountable.

That was a serious mistake. Floats are eminently countable, if we count
them by their representation size instead of using the inherent infinite
resolution sizes.
Now, that counting scheme certainly works flawlessly, but with any other
counting schemes, one side-effect is that in between any two items there
is zero "space" for an insertion. If there was a space and a third item
was inserted then that would mean that the counting scheme is flawed.
Let's say we count $100 bank notes : 1 $100, 2=$200, ... 10=$1000.
There is no way to find room for $139, as it is neither here nor there.
Yes, 100 < 139 < 200 is correct, but there is no corresponding counting
agent for it, as the resolution of value increments is $100.

```
To show that in our counting scheme there is no way to insert
another number, let's try to squeeze "0.333333000005"
between "0.333333" and "0.333334"
Please note that value-wise 0.333333 < 0.333333000005 < 0.333334 still
holds, so the insertion will work in that domain, but there is no gap
between the counting agents which we use when counting floats.
We can see below that a number "tmp_between" is actually very far away.
*);
tmp_low:=find_int_seqno_from_float_story ("0.333333")[1];
tmp_far_away:=find_int_seqno_from_float_story ("0.333333000005")[1];
tmp_high:=find_int_seqno_from_float_story ("0.333334")[1];
```

```
              tmp_low:=4800000

          tmp_far_away:=10199999700005

              tmp_high:=4800001
```

11 ######### Two way conversion of very large integers or floats

(* The floats we got from int_seqno in Maple-2018 or in Maple 2018
maxed out at 100 characters because some weird limitation.
Access into local lists in the a procedure failed when the list
length exceeded 100.

I re-engineered the code so that all access was restricted to
concatenating lists, avoiding random access altogether.
It took a lot of work as I had to comb through both conversion
procedures 7 and 8, but I was able to work with lists of over
50,000 elements when processing PI approximation of that length.
*);

```
first_float:=find_float_story_from_int_seqno(4^1167)[1];
first_integer:= find_int_seqno_from_float_story(first_float)[1];

second_float:=find_float_story_from_int_seqno(first_integer)[1]:
if (first_float = second_float)
  then print ("MATCH for two",Length(second_float),"digit floats");
end if;
second_integer:= find_int_seqno_from_float_story (second_float)[1]:
if (first_integer = second_integer)
  then print ("MATCH for two", evalf(log10(second_integer),5),
  "digit integers");
end if;
```

> *first_float* :=
>   "44439029053095321357741289673614347437938256318593647236788\
>   56393005768040355701356878902879710447137068341196632029 7689\
>   71652787366391513232434120214639952082192051311942302786 4000\
>   95877858318034513535955276225593705952946974289213162586 8915\
>   97808393674818485755386769605800262520764317746694821609 8658\
>   39216875599051584022873460784828167697230492986852805059 4896\
>   15469471904393458723030989210810077371099743406831417838 9821\
>   .93687734334763542623667903503431705173250369725811825869 716\
>   08551413713338747885841568992308301312590631470165561227 2473\
>   45551367711159904840973946624477562091197731535258896034 5461\
>   55638427508077276280600392614604818208785759232166308294 7134\
>   0745153688216274634533637323284089801568 71"

first_integer :=

    40179995126147785789221967160706252912694144430686734282513 1\
    09707537051912363201312211910125917394024233615070769688267 9\
    20744875086297523619091907081931759568739728461807480725077 6\
    00862900724862310621823597486030343535765227686029184632820\
    24380275543073366371798480926452202362686878859720253394488 7\
    92552951880391464256205861147063453509275074436881675245535 4\
    06539225247139541128507278902897290696339897690661482760550 8\
    39743189609012871883613011131530885346559253327532306432827 4\
    44769627234200487309725741209307747118331568323149005104522\
    61099623094004391435687655196202980588207795838173300643109 1\
    54007458475726954865254035335314433638790718330894967746524 2\
    06670638319394647171080273590955680821411 84

           "MATCH for two", 701, "digit floats"
           "MATCH for two", 702.60, "digit integers"


```
float_story2:=find_float_story_from_int_seqno(10^400)[1];
int_seqno2:= find_int_seqno_from_float_story(float_story2)[1];
Length(convert(float_story2, string));
```

float_story2 :=

    "4111111111111111111111111111111111111111111111111111111111 1\
    1111111111111111111111.11111111111111111111111111111111111111 1\
    11111111111111111111111111111111111111111111111111111111111 1\
    11111111111111111111111111111111111111111111111111111111111 1\
    11111111111111111111111111111111111111111111111111111111111 1\
    11111111111111111111111111111111111111111111111111111111111 1\
    1111111111111111111111111111111111111111"

int_seqno2 :=

    10000000000000000000000000000000000000000000000000000000000 0\
    00000000000000000000000000000000000000000000000000000000000 0\
    00000000000000000000000000000000000000000000000000000000000 0\
    00000000000000000000000000000000000000000000000000000000000 0\
    00000000000000000000000000000000000000000000000000000000000 0\
    00000000000000000000000000000000000000000000000000000000000 0\
    000000000000000000000000000000000000000000

399

```
# 12 ######### A strange pattern, in spite of the modulo arithmetic

float_story:="0.321";
for i from 1 to 4 do
  int_seqno:=find_int_seqno_from_float_story(float_story)[1];
  print("i=",i,"float_story=",float_story, "int_seqno=",int_seqno);
  float_story:=cat(float_story,"321");
end do;
```

$$float\_story := "0.321"$$

"i=", 1, "float_story=", "0.321", "int_seqno=", 2089

"i=", 2, "float_story=", "0.321321", "int_seqno=", 4789189

"i=", 3, "float_story=", "0.321321321", "int_seqno=", 7489189189

"i=", 4, "float_story=", "0.321321321321", "int_seqno=",
          10189189189189

```
> # 13 ######### Generating repeating int_seqno from repeating floats
float_rep_594739:="0.594739594739594739594739594739594739";
int_seqno:=find_int_seqno_from_float_story(float_rep_594739)[1];# repeats

float_rep_1919_1919:=
("1919191919191919191919.191919191919191919191919");
int_seqno1:=find_int_seqno_from_float_story (float_rep_1919_1919)[1];

float_rep_9191_1111:=("9191919191919191919191919.11111111111111111111111111");
int_seqno2:=find_int_seqno_from_float_story (float_rep_9191_1111)[1];

story_list:=["0",".","3","8","1"];
for i from 1 to 5 do
  story_list:=[op(story_list),"3","8","1"];
float_story:=Implode(story_list);
next_int_seqno:=find_int_seqno_from_float_story (float_story)[1];
print(float_story,next_int_seqno );
end do;

story_list:=["0",".","1"];
for i from 1 to 5 do
  story_list_01[nops(story_list)] := "0";
  story_list:=[op(story_list),"7"];
float_story:=Implode(story_list);
next_int_seqno:=find_int_seqno_from_float_story (float_story)[1];
print(float_story,next_int_seqno );
end do;
```

$$float\_rep\_594739 := "0.594739594739594739594739594739594739"$$

$$int\_seqno := 320352656352656352656352656352656635266$$

$$float\_rep\_1919\_1919 :=$$
$$"1919191919191919191919.191919191919191919191919"$$

$$int\_seqno1 := 20962727272727272727272727272727272727272728$$

$$float\_rep\_9191\_1111 :=$$
$$"9191919191919191919191919.11111111111111111111111111"$$

121

```
int_seqno2 := 2278727272727272727272727200000000000000000000000000
                                    ""
                 story_list := ["0", ".", "3", "8", "1"]
                        "0.381381", 4843243
                      "0.381381381", 7543243243
                    "0.381381381381", 10243243243243
                  "0.381381381381381", 12943243243243243
                "0.381381381381381381", 15643243243243243243
                                    ""
                     story_list := ["0", ".", "1"]
                           "0.17", 106
                          "0.177", 1960
                          "0.1777", 28600
                         "0.17777", 376000
                        "0.177777", 4660000
```

14 ######### List of odd patterns of converted repeating decimals

The output is periodic and it has identical repeating blocks as the input.
Discovery : The integer sequence numbers also have repeating blocks

```
f1_00001:="0.00001000010000100001000010000100001";
int_seqno:=find_int_seqno_from_float_story(f1_00001)[1];

f1_1234:="0.12341234123412341234123412341234"; #repeats 1234
int_seqno_f1_1234:=find_int_seqno_from_float_story(f1_1234)[1];

repeat_1_per_7 := cat("0",convert(evalf(1/7),string));
int_seqno_1_per_7:=find_int_seqno_from_float_story(repeat_1_per_7)[1];

repeat_1_per_11 := "0.09090909090909090909090909090909091"; #
int_seqno_1_per_11:=find_int_seqno_from_float_story(repeat_1_per_11)[1]; #

rep_1_per_17 := "0.05882352941176470588235294117647058823529";
int_seqno_rep_1_per_17:=find_int_seqno_from_float_story(rep_1_per_17)[1];

print("int_seqno_f1_per_17 blocks := 360 5294117647 0588235294117647
0588235294117");
```

f1_00001 := "0.00001000010000100001000010000100001"

int_seqno := 2610000900009000090000900009000090001

""

f1_1234 := "0.12341234123412341234123412341234"

int_seqno_f1_1234 := 2441107110711071107110711071107111

""

repeat_1_per_7 := "0.1428571429"

int_seqno_1_per_7 := 82285714287

""

repeat_1_per_11 := "0.09090909090909090909090909090909091"

int_seqno_1_per_11 := 2708181818181818181818181818181818182

""

repeat_1_per_17 := "0.05882352941176470588235294117647058823529"

int_seqno_repeat_1_per_17 :=
    3605294117647058823529411764705882352941177

"int_seqno_f1_per_17 blocks := 360 5294117647 0588235294117647
    0588235294117"

```
# 15 ######### Strange coincidence : int_seqnos = fake float_stories
#works only when the story's initial fake integer is larger than the
#int_seqno, e.g. "9.99" vs 990

for int_seqno from 990 to 1 by -1 do
  float_story:=find_float_story_from_int_seqno(int_seqno):# proc 8
  if float_story[2] = float_story[10] then
   print("int_seqno = " ,int_seqno,
     "float=",float_story[1]=",fake_float=",float_story[10]);
  end if;
end do;
```

```
              "int_seqno = ", 909, "float=", "9.09"=",fake_float=", 909
              "int_seqno = ", 908, "float=", "9.08"=",fake_float=", 908
              "int_seqno = ", 907, "float=", "9.07"=",fake_float=", 907
              "int_seqno = ", 906, "float=", "9.06"=",fake_float=", 906
              "int_seqno = ", 905, "float=", "9.05"=",fake_float=", 905
              "int_seqno = ", 904, "float=", "9.04"=",fake_float=", 904
              "int_seqno = ", 903, "float=", "9.03"=",fake_float=", 903
              "int_seqno = ", 902, "float=", "9.02"=",fake_float=", 902
              "int_seqno = ", 901, "float=", "9.01"=",fake_float=", 901
                "int_seqno = ", 9, "float=", "0.9"=",fake_float=", 9
                "int_seqno = ", 8, "float=", "0.8"=",fake_float=", 8
                "int_seqno = ", 7, "float=", "0.7"=", fake_float=", 7
                "int_seqno = ", 6, "float=", "0.6"=", fake_float=", 6
                "int_seqno = ", 5, "float=", "0.5"=", fake_float=", 5
                "int_seqno = ", 4, "float=", "0.4"=", fake_float=", 4
                "int_seqno = ", 3, "float=", "0.3"=", fake_float=", 3
                "int_seqno = ", 2, "float=", "0.2"=", fake_float=", 2
                "int_seqno = ", 1, "float=", "0.1"=", fake_float=", 1
```

```
# 16 ######### How to find the number of floats between any two floats

float_1:="0.01"; #
float_2:="0.002"; #
int_1:=find_int_seqno_from_float_story (float_1)[1];
int_2:=find_int_seqno_from_float_story (float_2)[1];
float_between_A := int_2-int_1+1;


float_1:="0.00000000000000000000000001"; #
float_2:="0.00000000000000000000000002"; #
int_1:=find_int_seqno_from_float_story (float_1)[1];
int_2:=find_int_seqno_from_float_story (float_2)[1];
float_between_B := int_2-int_1+1;

ratio_AB:=evalf(float_between_B/float_between_A);
```

$$\text{float\_1} := "0.01"$$

$$\text{float\_2} := "0.002"$$

$$\text{int\_1} := 91$$

$$\text{int\_2} := 1802$$

$$\text{float\_between\_A} := 1712$$

$$\text{float\_1} := "0.00000000000000000000000001"$$

$$\text{float\_2} := "0.00000000000000000000000002"$$

$$\text{int\_1} := 2340000000000000000000000001$$

$$\text{int\_2} := 2430000000000000000000000002$$

$$\text{float\_between\_B} := 2196000000000000000000000000002$$

$$\text{ratio\_AB} := 1.282710280 \; 10^{26}$$

```
# 17 ######### Integer output of Shifting decimal points to right

floats_in:=["1.23456789","12.3456789","123.456789","1234.56789",
"12345.6789","123456.789"]:
for i from 1 to nops(floats_in) do
 int_seqno:=find_int_seqno_from_float_story (floats_in[i])[1];
 print(floats_in[i], int_seqno);
end do;
```

$$\text{"1.23456789"}, 741111111$$

$$\text{"12.3456789"}, 1551111111$$

$$\text{"123.456789"}, 2361111111$$

$$\text{"1234.56789"}, 3171111111$$

$$\text{"12345.6789"}, 3981111111$$

$$\text{"123456.789"}, 4791111111$$

Conclusion, Footnotes

Widespread effect on current printed and online literature :

A game changer for theories on infinity, set theory and popular paradoxes etc.
Removes unsupportable unproven methods currently used to decide countability

Suggested changes :

Revision of entire body of literature, text books
Fix or remove internet publications
Current institutes teaching math curriculum
Expect near universal rejection, prepare for a long fight.

Practical Uses :

1. Gives an option to store floats as integers.
One bizarre byproduct of establishing universal and stationary one-to-one correspondence
between integers and floats is that now have the option to store floats as integers.
Then these "integer-looking" floats could be re-converted any time back to floats.

One very basic motivation to do such conversion would be the proliferation of float representation varieties.
For example we have scientific (mantissa, exponent) notation, decimal point alignment formatting issues
(e.g. 15.4e-20, 345.9889 and 0.0000075, then fractional or percent type floats. Integers have no such issues, alignment is trivial. so all the various floats would be represented by a single format.

Of course before using the integer number as a float again they would have to be converted back.
Luckily for us we did invent the two-way conversion algorithms.

2. Assign a place to a float in the counting hierarchy, that is permanently order each and every float in existence.

3. Obtain distance between any two floats or between floats given any two counting integers.
4. Of course debunking the entire body of false theories about relative cardinality of floats and integers.

5. Disproves Cantor's Diagonal Argument and all derivative assumptions.
In essence all complicated, convoluted and "in-your-face" arguments, insults and derogatory \
name-calling melt away into pitiful irrelevance. Whatever CDA concludes it must exclude the
claim that floats are impossible to count. We are eminently capable of doing it as I type this.

6. But CDA does actually PROVE something incredibly valuable. If you repeat some idea over and over and
posses charm, coupled with the ability to persuade, you can sell virtual and actual BS no matter how
outrageous it might be. This goes for dictators, religions (mainstream and cult), psychiatry, diets, drugs,
exercise fads. Let's not forget the long vicious, bloody and ultimately genocidal history of acquiring
then maintaining power over the ages.

7. Demolished all set theories about relative countability of various classes of floats,
(such as rational floats are more countable than the transcendental numbers).

8. Allows uniform treatment of all floats regardless of appearance or special status.

9. Establishes and demonstrates a clear demarcation line between algorithms and floats.

10. Removes the exalted and mysterious status accorded to the manipulation of floats.
11. Created strict, sufficient syntax rules for integer and floats

Add on the bottom : temporary tagging. of toys
Next counting would result in new tags, so they would not be a good way to identify toys.

Footnotes :

1. When counting floats we will make sure that we count every one of them and do so only exactly one time. Depending on the generating algorithms we will either ignore duplicates, or won't generate them at all in the first place.

2. The act of orderly counting valid floats qualifies as an integer algorithm which produces each and every float starting from 0.1 to an arbitrarily large value. We can think of this process as the "painting" of the number line in a systematical fashion, where with each round both the resolution(or representation length) and the range of the floats is incremented.

3. When we execute the generating algorithm which returns an integer sequence number associated with a float F, there is absolute guarantee that the algorithm will compute the exact number of all floats preceding F.

4. It is this ability of the algorithm which obliterates 6 generations of completely misguided and worthless theorems claiming that the floats cannot be counted. From now on each and every float - no matter how large - can be counted and accounted for. By accounted for we mean that as we speak each float has its very own unalienable integer sequence number forever and after .

5. One might be tempted to compare the unique and permanent pairing of integers and floats to the US Social Security numbering system (#SS) . Fair enough but not good enough :

5.1   There is absolutely nothing inherent in the SS # 123-45-6789 which would lead to John Smith 2. Same goes for the reverse : there could be any number of John Smiths, and in the absence of additional corroborating data, we can't retrieve his SS#

5.2   The only way the system works is by utilizing a central SS# database which will store the SS# - person association. Once that database gets corrupted or blown away the whole system becomes useless. Everybody could claim any SS# because nobody would be able to challenge him.

5.3 SS# are re-used after the person who had it passes away.

5.4 In contrast :
1. For a valid set of syntax rules the float ←→ integer relationship lasts forever and cannot be corrupted
2. You may retrieve a float from known integer and vice-versa, without an intermediate database.
3. Neither floats nor integers disappear or die off !!!

Appendix

Case Study #1 : Converging geometric series using ω instead of ∞

Probably the best known and universally accepted algorithm is the sum of converging geometric series. There is a well known formula for the case when the number of terms is finite.

The unwarranted extension modified this formula and renamed it the "sum of infinite series" We should be able to remove infinity from all derivations. I will demonstrate this on the sum of geometric series with a positive ratio of $q < 1$

Let $\mathbf{k}$ be a positive integer and $\mathbf{q = 1/k}$ a fraction , where ω can grow without bounds.

Then the sum of $\mathbf{1}$ and $\boldsymbol{\omega}\text{-}\mathbf{1}$ fractions from $\mathbf{q}$ to $\mathbf{q}^{\omega-1}$ is :

$$\text{SUM}(\omega, q) := 1 + q + q^2 + \dots + q^{\omega-1}$$

or

$$\text{SUM}(\omega, q) := \frac{1 - q^\omega}{1 - q}$$

Since $\mathbf{q < 1}$ then the value of $\mathbf{q}^\omega$ is approaching 0 as $\boldsymbol{\omega}$ increases. Without making any assumptions we arbitrarily replace $\mathbf{q}^\omega$ with 0 in the $\text{SUM}(\omega, q)$ formula.

NO, we are not setting $\mathbf{q}^\omega$ to zero! Instead we **mutilate** the SUM formula by removing the term $\mathbf{q}^\omega$ and sticking 0 in place of it.

YES, we are doing it on a whim to see what we end up with.

Let's name what we actually removed from the SUM formula a GAP, which gives us :

$$\text{GAP}(\omega, q) := \frac{q^\omega}{1 - q}$$

Finally what remains of the SUM formula is obviously the LIMIT : Then we have :

$$\text{LIMIT}(q) := \frac{1}{1 - q}$$

Last but not least is the amount of advance or JUMP from one SUM to the next :

$$\text{JUMP}(\omega, q) := \frac{q^{\omega+1} - q^\omega}{q - 1}$$

$$\text{JUMP}(\omega, q) := q^\omega$$

We can write the relationship between the 4 pivotal formulas :

$$\text{SUM}(\omega, q) := \text{LIMIT}(q) - \text{GAP}(\omega, q)$$
$$\text{JUMP}(\omega, q) := \text{SUM}(\omega, q + 1) - \text{SUM}(\omega, q)$$
$$\text{LIMIT}(q) := \text{SUM}(\omega, q) + \text{GAP}(\omega, q)$$
$$\text{GAP}(\omega, q) := \text{LIMIT}(q) - \text{SUM}(\omega, q)$$

5.1 Example : Let $q = 1/2$ and $\omega = 4$ then the sum of fractions from $n = 1$ to 4 is :

$$\text{SUM}(\omega, q) := 1 + q + q^2 + \ldots + q^{\omega - 1} \quad = 1 + \tfrac{1}{2} + \tfrac{1}{4} + 1/8 + 1/16 = 30/16 = 15/8$$

$$\text{SUM}\left(4, \frac{1}{2}\right) := \frac{15}{8}$$

$$\text{JUMP} := \frac{1}{16}$$

$$\text{GAP}\left(4, \frac{1}{2}\right) := \frac{1}{8}$$

$$\text{LIMIT}\left(\frac{1}{2}\right) := 2$$

5.2 Example : Let $q = 1/2$ and $\omega = 100$ then the sum of fractions from $n = 1$ to 100 is :

$$SUM\left(100, \frac{1}{2}\right) := \frac{1267650600228229401496703205375}{633825300114114700748351602688}$$

$$LIMIT\left(\frac{1}{2}\right) := 2$$

$$GAP\left(100, \frac{1}{2}\right) := \frac{1}{633825300114114700748351602688}$$

$$q := \frac{1}{2}$$

5.3 Example : Let q = 1/2 and ω range from 1 to 5 . We show the relationship between the JUMP and the GAP parameters :

omega ω	SUM	GAP	JUMP	JUMP DIVIDED BY GAP	GAP MINUS JUMP	LIMIT
1	1	1	1/2	1/2	1/2	2
2	3/2	1/2	1/4	1/2	1/4	2
3	7/4	1/4	1/8	1/2	1/8	2
4	15/8	1/8	1/16	1/2	1/16	2
5	31/16	1/16	1/32	1/2	1/32	2

Note that in all three cases LIMIT is not dependent on ω and it evaluates to 2 if q = 1/2
AS ω grows, the SUM grows, the GAP narrows and approaches 0. However for no ω value
will it ever reach zero. The LIMIT remains stationary since it is not a function of ω anymore.

It is imperative to maintain the distinction from now on : The quantity computed by the LIMIT formula
is a constant and it has absolutely nothing to do with summing fractions. You can regard the LIMIT
as a butchered, neutered, crippled SUM which turned into a frozen piece of gatepost.
It simply stands as a guard or a pie-in-the-sky wish for those who compute the ever increasing quantity
SUM (ω,q) . If they ever come up with a value for SUM which exceeds the LIMIT they will be alerted
that they made a mistake. If no alarm is sounded it does not mean that their computations are correct.

Note that we introduced ω with the express purpose and stipulation that it must remain a number
forever and can never be re-assigned to any other mathematical entity (yes, I must mention the
dreaded word, ω cannot become infinity, by the virtue of defining it as a valid, computable number.

1. I strongly suggest that the designation "infinite sums or sums of infinite series" be phased out,
for the inescapable reason that there is no such a thing. It is not possible to calculate infinite sums.
We must use the formula SUM(ω,q) which will be forever an approximation and LIMIT (q)
to denote a barrier, but not a sum.

2. Here is what Wikipedia Geometric_series article and I am sure most other tutorials call the sum
of infinite series :

"As n goes to infinity, the absolute value of r must be less than one for the series to converge.
The sum then becomes 1/(1-r) (or in my notation 1/(1-q)"

Wrong !!! That formula is definitely not the sum of the series but the LIMIT of it as I described
above. As a matter of fact the usual formulas which compute partial sums for ω will never produce
the universally accepted infinite sum (which we now call a LIMIT) That limit was reached and
computed by an act of aggression visited on the SUM formula by human agents.

3. Those people who attempt to reach LIMIT by hopelessly re-computing SUM(ω,q) will be
laboring away when the last electron of the Universe blinks out. The elusive limit will always
be just around the corner.

4. What I call the LIMIT is a result of a decidedly non-mathematical act, which changed a mathematical progression of sums into a single static value and by doing it changed its nature : It is no longer a changing sum but a frozen goalpost. As such the two are grossly incompatible and the nomenclature should reflect it.

Instead, the designation of SUM was retained when no summing was taking place anymore (there is no place in the LIMIT expression to plug in values as it is only a function of q). In addition an even higher order transgression was committed. Instead of using a legitimate variable ω a non-existent and non-reachable figment of someone's imagination the mystical "infinity" was invoked. Sound mathematics receded in the rear-view mirror ever since with unfortunate consequences ever since.

5. **The LIMIT is never part of the series and the series can never reach the LIMIT value.**
The two twain shall never meet. Of course you can call it whatever you like, but after you read this at least you will be aware that you are doing it wrong. Need I say more ?
7. ω is tangible but infinity is not even imaginary. When the jump is made from ω to infinity in order to draw conclusions there is always an act of violence, a mutilation involved.

8.It is usually the formula for sums which is defaced, in order to get the limit. That of course is a legitimate hammer in our toolbox, we just have to make sure that once we clobbered the beast we stop calling it a **"sum of series"** and definitively never, ever call it the **sum of infinite series**. It has a respectable and correct name : it is called the **limit of convergence of the series.**

9. We were able to define sums which could involve an arbitrarily big number of elements.

10. Without using infinitesimal arithmetic we easily extracted the formula for LIMIT which is the smallest quantity the geometric series will never reach.

11. What is somewhat remarkable, that in the case of geometric series with an integer ratio of $0 < q < 1$ we know the hard limit of the SUM as soon as q is known.
(We are at liberty to define any other "sloppy" limit L as long as L > LIMIT.)

12. It is my contention that simply by replacing the various convoluted gymnastics (which involve the baggage of infinity without offering any benefits) we could survive with a less kinky and far more tangible construct the ω.

13. Very important to note that the LIMIT(q), GAP(ω,q) and SUM(ω,q) formulas are independent from each other as each is the function of either ω and q or q only.

11. Zeno's paradox and how to resolve it.

This anecdote is not a paradox at all, it is jut a story made contradictory and subsequently nonsensical by ignoring or altering some physical facts. What fact do I talk about. It is that any physical entity possesses finite sizes, in our world the limit is electrons, protons and neutrons. Assume that Zeno was a walking hydrogen atom with 2 electrons for legs and he advanced by always putting a heel right next ahead of the other toe. Then his advances would have the minimum resolution of one electron size. It is easy to see that he would always cross the finish. No paradox there.

Now let's just forget the tiny shoes, and imagine the following generic scenario:

1. We have any task ahead of us

2. We have the ability to define a subset of this task
3 . Complete only that subset of the task
4 . Define the remaining task as the "task ahead of us"
5. Repeat steps 2, 3, 4

If we do that then we will never finish the task. No paradox there either.
The general principles of advancing toward a limit are : (this needs work !!!)

A : There are three ways we will never reach the limit :
1. We don't advance
2. We measure the distance but always advance the fraction of that distance.
3. The limit moves away from us

B: There are 3 ways we cannot avoid reaching the limit :
1. We have no idea what the distance is and we eventually breach the limit
2. We do know what the exact distance to the limit is, however due to
   physical or mental constraints we are unable to advance less than that distance
   (this would be Zeno with the electrons for legs)
3. The limit moves toward us

C: In converging series as in the converging geometric sum, the limit will not be reached.
    This is guaranteed by the design of the algorithm which generates subsequent members
    of the series.
    Once such property of the series is verified and the limit is computed we can trust the
    algorithm, that is the series is put on an autopilot. Usually we can derive the algorithm
    which will produce the gap between the current value of the series and the limit.

Case Study #2: Disproving Cantor's Diagonal Argument

It was simply not considered worthy of becoming a target of rigorous analysis. As I will show its detractors were right. The CDA is outright gibberish and in the following pages I will present my case.

2.1 We can disprove CDA for the total absence of floating point considerations (uses syntax of .0, but never uses floating point numerical properties. The original used binary, the contemporary version switched to base 10)

2.2 The CDA always uses notation for infinity but completely ignores the actual variation set sizes for finite float lengths (uses 6 characters, displays 10 rows out of a million possible)

2.3 It fails to work for any concrete length floating point strings less than infinity in size which of course is all the floats we can ever access. (We use ω)

2.4 It fails to prove that the diagonal float created is not a member of the generation of variations even for the sample size which used in the argument (uses 6 digit decimal diagonal, which is clearly is a member of a 1000000 strong variation provided that the symbol sets match.)

3. Show that the argument actually applies to integers only :

The left column of the table contains a list of integer counting numbers, while the center column has a list of an unlimited number of floats whose character count is also unlimited.

The rightmost column duplicates the center column with a single digit bolded in a pattern which progresses to the right and down, thus forming a diagonal float.

The ellipses (**. . .**) indicate unlimited and unknown extent presumably all the way to infinity.

Table 1 CDA Applied to Fake Floats (as it is usually presented today)

Integer counting numbers	Floats	Diagonal marked
1	0.799999...	0.**7**99999...
2	0.000038...	0.0**0**0038...
3	0.116921...	0.11**6**921...
4	0.138443...	0.138**4**43...
5	0.141579...	0.1415**7**9...
6	0.495985...	0.49598**5**...
...	...	...

We only show six floats and our example diagonal float also shows 6 fractional digits

I must mention here that the possible number of unique 6-digit fractional floats is actually 999999, so this table is grossly misleading .

Now we extract the marked digits from the third column to form a diagonal float.
The first 6 digits of the diagonal float are : **0.706475 ...**
We alter each fractional digit to obtain a new float : **0.817586 ...**

Please note : Any digit different from the original works. (Here, I incremented each digit by 1)

Cantor has claimed that this recently generated float from the diagonals is
not found in the list, therefore it is not possible to count every one of them.

He then concludes for some incomprehensible reason that there are no more integer
counting numbers to match this recently composed real number.

Why not ? After all, the set of counting numbers are arbitrarily large so we can
always take the next one to count whatever new real number anybody composed.
He never explains why and when our ability to generate counting integers have been lost.

Oddly, there is absolutely no manipulation of the floats in the center column
which would indicate that they are nothing more than a sequence of integer digits.

The negative powers of 10 (or any base) are completely irrelevant and never
invoked in any computations.

Thus we are altogether justified if we remove the two-character prefix ".0"
or "zero-dot", the decimal point and the digit zero.

4. CDA applied to a finite set of integers :

Instead of infinite extents we use all possible D=6 digit variations of the
10-digit decimal integer symbol set [0,1,2,3,4,5,6,7,8,9] so S=10

Note: To count D-element variations of S objects, we first need to choose
a D-element combination and then permutate the selected objects.

This will generate combinations with leading zeros, which we will allow.

Table 2 CDA Applied to a finite set of integers

Integer counting Numbers (ICN)	Integers	Diagonal marked
1	000000	**0**00000
2	000001	0**0**0001
3	000002	00**0**002
4	000003	000**0**03
5	000004	0000**0**4
6	000005	00000**5**
. . .	. . .	end
999999	999998	
NUM_ICNS := S^D 1000000	LAST_INT := $S^D - 1$ 999999	

The 6 digits of the diagonal integer are : **0 0 0 0 0 5**
Alter each digit randomly to obtain a new integer : **8 1 7 5 8 0**

Neither this new integer nor any other six-digit decimal integer can be novel :
that is it must already exists in the list of "Integers" in the center column.

Proof : Follows by the rules of generating all the possible variations
of the 10-symbol decimal set.

Then any D-digit number (which consists only one or more of the S symbols)
will have to be one of the S^D variations (ranging from 000000 to $S^D - 1$)

5. CDA applied to bare-bone binary integers

Table 3 CDA Applied to 3-digit Binary Numbers

Integer counting numbers	Binary Digits	Diagonal marked
1	000	**000**
2	001	**001**
3	010	**010**
4	011	The End
5	100	
6	101	new binary
7	110	integer found
8	**111**	**111**

The 3 digits of the diagonal integer are :   **0 0 0**
Altered (inverted) each digit to obtain a new binary integer : **1 1 1**

This is an example where all available variations are displayed so we could
locate the new binary integer in the list of all possible 3-digit integers

Table 4 CDA Applied to 4-digit Binary Numbers

Integer counting numbers	Binary Digits	Diagonal marked
1	0000	**0000**
2	0001	**0001**
3	0010	**0010**
4	0011	**0011**
5	0100	The End
6	0101	
7	0110	
8	0111	
9	1000	
10	1001	
11	1010	new binary
12	1011	integer found
13	**1100**	**1 1 0 0**
14	1101	
15	1110	
16	1111	

The 4 digits of the diagonal integer are :   **0 0 1 1**
Altered (inverted) each digit to obtain a new binary integer : **1 1 0 0**
Once again all available variations are displayed so we can locate the new
integer in the list of all possible 4-digit binary integers

6. Conclusion

6.1 The mechanism of generating new integers is demonstrated first for the 3-digit case.
6.2 We show that the generated integer will exist in the list of all 3-digit integers.
6.3 An identical procedure for the 4-digit integers yielded the same results.

> 6.4 If a procedure works for a positive integer n=k and we prove it for n = k+1,
> we expect it to work for any integer ω, no matter how big. (I must avoid saying
> that it would work for infinity because we cannot use infinity in computations.)
> This is the well-known proof by induction.

6.5 To disprove this chain of reasoning one must show a possible mechanism which
      will terminate or invalidate the procedure starting at some very big number or ω.

6.6 The steps presented for integers apply both to fractional only and full-blown floats.

6.7 For fractional only floats we simply prefix 2 inert characters "dot-zero" or ".0" and
      we are done. Those two characters are props only and server no function whatsoever.
      So the proof applies to the **0.f** subset of floats by analogy.

6.8 The argument could be extended to full-blown legitimate syntax floats like **w.f**, where
      w stands for an integer representing whole numbers, and **f** indicates a reversed order
      integer representing the fractional part. However, we will not bother. CDA has been
      disproved to my own satisfaction for the binary case.

6.9 The odd diagonal generation trick has survived for 120 years, we can finally retire it.

6.10 Amazingly, we have had in plain view a far more powerful method of producing a set
      of numbers whose each member is guaranteed to differ in at least one digit : not only
      from a subset but from all the numbers comprising the rest of the set.

We accomplish this when we generate the **variation** of a set comprising unique symbols.

7. Notes on the History of CDA and Counting

In the history of mathematics literature there was some mention that Cantor's contemporaries
really disliked the diagonal argument, calling it metaphysical or such. (check that).

If you are a mathematician and come up on a theorem, you attempt to prove or disprove them.
Once there is a proof, liking or detesting is an irrelevant response by a professional.
Cantor's "diagonal proof" was not amenable for rigorous logical treatment, which is a good
indication that it was outside of or contrary to the accumulated body of mathematical knowledge
to be even considered for a worthy counterattack by rigorous analysis. It is bordering on gibberish.

See the references for quotes and articles on this subject !!!

References

Cantor's paradise is an expression used by <u>David Hilbert</u> (<u>1926</u>, page 170) in describing <u>set theory</u> and infinite <u>cardinal numbers</u> developed by <u>Georg Cantor</u>.

Floating Point to Integer Mapping (related in name, vastly different in scope)
by Motion Imagery Standards Board 2/27/2014
http://www.gwg.nga.mil/misb/docs/standards/ST1201.1.pdf

That is it, see http://people.csail.mit.edu/alinush/math/countability.pdf for details

1. Attempts to prove that $1+2+3+..+ =-1/12$
ASTOUNDING: $1 + 2 + 3 + 4 + 5 + ... = -1/12$
 https://www.youtube.com/watch?v=w-I6XTVZXww&t=22s
The original video as far as I can tell

2. Video debunking the ASTOUNDING… video
Numberphile v. Math: the truth about 1+2+3+...=-1/12
https://www.youtube.com/watch?v=YuIIjLr6vUA

3. Why -1/12 is a gold nugget (I know, because YouTube pays for it)
https://www.youtube.com/watch?v=0Oazb7IWzbA

4. What do we get if we sum all the natural numbers? By Dr. Tony Padilla
https://www.nottingham.ac.uk/~ppzap4/response.html

*"There is an enduring debate about **how far we should deviate from the rigorous academic approach in order to engage the wider public.** From what I can tell, our video has engaged huge numbers of people, with and without mathematical backgrounds, and got them debating divergent sums in internet forums and in the office. That cannot be a bad thing and I'm sure the simplicity of the presentation contributed enormously to that. In fact, if I may return to the original question, "what do we get if we sum the natural numbers?", I think another answer might be the following: we get people talking about Mathematics." By Dr. Tony Padilla*

My comment : You get them involved, present material which is total nonsense and turn them off to math for the rest of their lives, thinking that they are grossly outclassed and will never understand it . [TMV]

Cesàro summation assigns values to some <u>infinite sums</u> that are <u>not convergent</u>
https://en.wikipedia.org/wiki/Ces%C3%A0ro_summation

Example treatment of Cantor's Diagonalization Proof
 http://people.csail.mit.edu/alinush/math/countability.pdf for details

Today, a few serious mathematicians still refute CDA [<u>Norman Wildberger</u>]
https://www.youtube.com/watch?v=XKy_VTBq0yk

Anecdotal references related to CDA

Paul du Bois-Reymond discovered a proof method that later became
known as the Cantor's diagonal argument.[1]   So the argument is Cantor's
only by possession and name only, but not by right.

Cantor, a devout Lutheran,[4] believed the theory had been communicated
to him by God. (always a good last resort line to invoke divine authority
when the facts are not on your side.)

Henri Poincaré referred to his ideas as a "grave disease" infecting the
discipline of mathematics,[8] https://bonald.wordpress.com/2017/11/03/corrupter-of-youth/
He said : "Most of the ideas of Cantorian set theory should be banished from mathematics
once and for all."

Leopold Kronecker's public opposition and personal attacks included describing
Cantor as a "scientific charlatan", a "renegade" and a "corrupter of youth."
https://bonald.wordpress.com/2017/11/03/corrupter-of-youth/

Note : (Instead of insults, why didn't these two just prove that the argument was invalid ?)

Leopold Kronecker claimed: "I don't know what predominates in Cantor's theory –
philosophy or theology, but I am sure that there is no mathematics there"
https://en.wikipedia.org/wiki/Controversy_over_Cantor%27s_theory

"Actual infinity does not exist. What we call infinite is only the endless possibility
of creating new objects no matter how many exist already". (Poincaré quoted from Kline 1982)
https://en.wikipedia.org/wiki/Controversy_over_Cantor%27s_theory

Carl Friedrich Gauss's views on the subject can be paraphrased as: 'Infinity is nothing more than a
figure of speech which helps us talk about limits. The notion of a completed infinity doesn't belong
in mathematics'. In other words, the only access we have to the infinite is through the notion
of limits, and hence, we must not treat infinite sets as if they have an existence exactly
comparable to the existence of finite sets.

https://en.wikipedia.org/wiki/Controversy_over_Cantor%27s_theory

===========================

For Chapter 3a :
 https://en.wikipedia.org/wiki/Real_number algorithms vs. real numbers

Cardinality of sets and Cantor:
https://math.berkeley.edu/~arash/55/2_5.pdf   Berkeley

http://mathforum.org/library/drmath/view/52393.html
proof that there are more reals between 0 and 1 than there are integers

Google hits

https://stackoverflow.com/questions/19254355/is-set-of-real-numbers-between-0-and-1-really-
uncountably-infinite

Uncountable sets :
 http://www.math.umaine.edu/~farlow/sec25.pdf

Has an overall description of set theory
http://www.msc.uky.edu/droyster/courses/fall06/PDFs/Chapter03.pdf

Has list of people who defied Cantor, but also has some contemporary
nonsense about the set of reals. Also has the 3x3 matrix of photos and quotes
http://www.science4all.org/article/cantors-infinite/

I have found two references for complete trees :
http://cs-study.blogspot.com/2012/11/complete-binary-tree.html

 https://hbfs.wordpress.com/2009/04/07/compact-tree-storage/

http://steve-patterson.com/cantor-wrong-no-infinite-sets/

[https://www.scientificamerican.com/article/strange-but-true-infinity-comes-in-different-sizes/]

http://people.csail.mit.edu/alinush/math/countability.pdf

Topic : There is a claim that the set of real numbers is bigger than the set of integers.

This claim was based on the mistaken belief (no, dogma) that because we can never line up
 all the real numbers in a particular order, we cannot establish a one-to-one correspondence between the
set of real numbers and the set of natural numbers, and therefore the cardinality of the set of real
numbers cannot be $\aleph_0$. The cardinality of the set of real numbers must be a different infinite "number!"

"Another Infinite "Number": $\aleph_1$
Clearly $\aleph_1$ must be a bigger infinity than $\aleph_0$, because the set of real numbers is a bigger set than the set
of natural numbers. Just like we say that something is COUNTABLE if it has a cardinality of $\aleph_0$, we say
that a set is UNCOUNTABLE if it had cardinality $\aleph_1$. We describe sets with the cardinality $\aleph_1$ as
UNCOUNTABLE because they cannot be lined up in order and therefore "counted." If these sets could
be lined up in order and "counted," we would be able to create a one-to-one correspondence with the
integers, and that would mean the set had cardinality $\aleph_0$ instead of $\aleph_1$."

 http://www.cwladis.com/math100/Lecture5Sets.htm **[have it in file "cardinality of sets"]**

Topics related to Cantor's Diagonal Arguments, set theory, uncountability of sets…

!!!!!!!! I guess that ship has sailed !!!!!!!!!